__Disclaimer__

The publisher of this book is by no way associated with the National Institute of Standards and Technology (NIST). The NIST did not publish this book. It was published by 50 page publications under the public domain license.

50 Page Publications.

Book Title: Evaluating Reasoning Systems

Book Author: Conrad E. Bock; Michael Gruninger; Don E. Libes; Joshua Lubell; Eswaran Subrahmanian

Book Abstract: A review of the literature on evaluating reasoning systems reveals that it is a very broad area with wide variation in depth and breadth of research on metrics and tests. Consolidation is hampered by nonstandard terminology, differing methodologies, scattered application domains, unpublished algorithmic details, and the effects of domain content and context on the choice of metric and tests. The field of information metrology, which applies to reasoning as a kind of information processing, is still emerging from ad hoc experience in evaluating narrow kinds of information systems. This report begins to bring order to the area by categorizing reasoning systems according to their capabilities. The characteristics of each category can be used as a basis for evaluating and testing reasoning systems claiming to be in that category. Capabilities are analyzed along several dimensions, including representation languages, inference, and user and software interfaces. The report groups representation languages by their relation to first-order logic, and model-theoretic properties, such as soundness and completeness. Inference procedures are divided into deduction, induction, abduction, and analogical reasoning. Capabilities of user and software interfaces are described as they apply to reasoning systems. The report introduces information metrology, model theory, and inference to facilitate understanding of the reasoning categories presented. It concludes with recommendations for future work.

Citation: NIST Interagency/Internal Report (NISTIR) - 7310

Keyword: reasoning categories;reasoning systems;software metrics

NISTIR 7310

Evaluating Reasoning Systems

Conrad Bock
Michael Gruninger
Don Libes
Joshua Lubell
Eswaran Subrahmanian

National Institute of Standards and Technology
Technology Administration, U.S. Department of Commerce

NISTIR 7310

Evaluating Reasoning Systems

Conrad Bock
Michael Gruninger[1]
Don Libes
Joshua Lubell
Eswaran Subrahmanian[2]
Manufacturing Systems Integration Division
Manufacturing Engineering Laboratory

[1] Also at the University of Toronto, and during some of the preparation of this report, at the University of Maryland, College Park.

[2] Also at Carnegie Mellon University.

May 2006

U.S. DEPARTMENT OF COMMERCE
Carlos M. Gutierrez, Secretary
TECHNOLOGY ADMINISTRATION
Robert Cresanti, Under Secretary of Commerce for Technology
NATIONAL INSTITUTE OF STANDARDS AND TECHNOLOGY
William Jeffrey, Director

Executive Summary

A review of the literature on evaluating reasoning systems reveals that it is a very broad area with wide variation in depth and breadth of research on metrics and tests. Consolidation is hampered by nonstandard terminology, differing methodologies, scattered application domains, unpublished algorithmic details, and the effects of domain content and context on the choice of metric and tests. The field of information metrology, which applies to reasoning as a kind of information processing, is still emerging from ad hoc experience in evaluating narrow kinds of information systems.

This report begins to bring order to the area by categorizing reasoning systems according to their capabilities. The characteristics of each category can be used as a basis for evaluating and testing reasoning systems claiming to be in that category. Capabilities are analyzed along several dimensions, including representation languages, inference, and user and software interfaces. The report groups representation languages by their relation to first-order logic, and model-theoretic properties, such as soundness and completeness. Inference procedures are divided into deduction, induction, abduction, and analogical reasoning. Capabilities of user and software interfaces are described as they apply to reasoning systems. The report introduces information metrology, model theory, and inference to facilitate understanding of the reasoning categories presented. It concludes with recommendations for future work.

Applying the results of this report to evaluation of specific reasoning systems requires further refinement of reasoning categories as needed to support development of test cases of interest to reasoning system users. Then development can proceed on generic test cases for reasoning categories. These are independent of the application, reasoning tool, and computational platform. Finally, based on the generic tests, development can start on specific test cases that are dependent on application, reasoning system tool, and computational platform. These steps can also be applied to user interface and software interface capabilities. This work can only be completed in partnership with users of reasoning systems and reasoning tool providers, to focus effort and reach the level of completeness necessary to put these metrics into practice.

Table of Contents

1 Introduction..5
2 Information Metrology..5
3 Reasoning..9
   3.1 Introduction to Reasoning..10
     3.1.1 Model Theory..10
     3.1.2 Inference...13
   3.2 Representation Languages...15
     3.2.1 First-Order Logic..16
     3.2.2 Restrictions of First-Order Logic..17
       3.2.2.1 Monadic First-Order Logic..17
       3.2.2.2 Universal-Existential Sentences..17
       3.2.2.3 Universal Sentences..17
       3.2.2.4 Horn Clauses..18
       3.2.2.5 Datalog...19
       3.2.2.6 Existential Sentences...19
     3.2.3 Beyond First-Order Logic...19
       3.2.3.1 Transitive Closure Logic...19
       3.2.3.2 Least Fixpoint Logic..19
       3.2.3.3 Partial Fixpoint Logic..20
       3.2.3.4 Second-Order Logic..20
       3.2.3.5 Monadic Second-Order Logic...20
       3.2.3.6 Infinitary Logic..21
       3.2.3.7 CycL...21
     3.2.4 Reified First-Order Logics..22
       3.2.4.1 Common Logic...22
     3.2.5 Description Logics..23
     3.2.6 Web Languages..24
       3.2.6.1 RDF/S..24
       3.2.6.2 OWL...25
     3.2.7 Modal Logics..26
       3.2.7.1 Propositional Modal Logics...26
       3.2.7.2 Quantified Modal Logics...28
     3.2.8 Nonmonotonic Logics...28
       3.2.8.1 Reiter's Default Logic..28
       3.2.8.2 Model Preference Default Logic...29
       3.2.8.3 Circumscription...29
       3.2.8.4 Autoepistemic Logic..30
     3.2.9 Additional Terminology..30
   3.3 Inference..32
     3.3.1 Deduction..33
       3.3.1.1 First order deduction..33
       3.3.1.2 Rule Systems..36
         3.3.1.2.1 Intersection of Inference and Programming..............................36
         3.3.1.2.2 Basic Rule System Semantics..38

   3.3.1.2.3 Rule System Inference..40
   3.3.1.2.4 Rule System Semantics Continued..44
  3.3.1.3 Temporal Reasoning..53
   3.3.1.3.1 Types of Temporal Reasoning System...53
   3.3.1.3.2 Evaluation of temporal systems..54
 3.3.2 Induction..55
  3.3.2.1 Definitions of Induction...56
  3.3.2.2 Measurement for Machine Learning Systems..................................57
   3.3.2.2.1 Evaluating Machine learning systems...57
   3.3.2.2.2 Rule Quality Measures for Rule Induction Systems..........................60
 3.3.3 Abduction...62
  3.3.3.1 Best explanation Model of Abduction..63
   3.3.3.1.1 Josephson best explanation model..63
   3.3.3.1.2 Konolige Model for Abduction...63
   3.3.3.1.3 Limitations of the Best Explanation Models of Abduction................65
  3.3.3.2 Creative/selective Model..66
  3.3.3.3 Conclusions on Abduction...67
 3.3.4 Analogical reasoning..68
  3.3.4.1 Structure Mapping...68
  3.3.4.2 Copy-Cat Model of Analogical Reasoning and Its Extensions.................69
  3.3.4.3 Hybrid Model of Analogy..71
4 User Interfaces..71
 4.1 User Interface Types...71
 4.2 Front-end Methods and Metrics...72
 4.3 Approaches..72
 4.4 Back-end Methods and Metrics...72
 4.5 Physical Methods...73
 4.6 Products and Toolkits...73
 4.7 Representations and Standards..74
 4.8 Comparisons..74
5 Software Interfaces...74
 5.1 Methods of Interaction..75
  5.1.1 Data Exchange...75
  5.1.2 Application Program Interfaces (APIs)...76
 5.2 The Value of Standards..76
 5.3 Software Interface Quality..78
6 Conclusion and Future Work..78
7 Acknowledgements and Disclaimer..79
8 References..79

1 Introduction

This report provides a basis for evaluating and testing reasoning systems by categorizing them according to their capabilities. These categories give metrics for determining if a particular system is performing the required kinds of reasoning. The capabilities are analyzed along several dimensions, including representation languages, inference, and user and software interfaces. It groups representation languages by their relation to first-order logic, and model-theoretic properties, such as soundness and completeness. Inference procedures are divided into deduction, induction, abduction, and analogical reasoning. Capabilities of user and software interfaces are described as they apply to reasoning systems. The report introduces information metrology, model theory, and inference to facilitate understanding of the reasoning categories presented.

Section 2 discusses the state of information metrology in general, providing an initial framework for analysis of reasoning systems. Section 3 covers reasoning, both representation and inference (items 2a and 2b). Section 4 addresses user interfaces (item 2c), and Section 5 software interfaces (item 2d). Section 6 gives conclusions and future work.

2 Information Metrology

The study of measurement (*metrology*) has been primarily applied to physical artifacts. The U.S. National Institute of Standards and Technology (NIST) is well known for its work in this area. For example, NIST maintains the national standard for mass in the form of a prototype kilogram and provides services to the nation in mass calibration [NIST 2005a]. It provides similar services for other units, such as length calibrating against the standard definition of the meter, and calibration for many other units, such as angles, pressure, luminosity, and voltage. Increasingly precise measurements of these units are essential to U.S. commerce as the economy develops more sophisticated machinery, and uses larger quantities of the entities being measured.

Generalizing from the above experience, we can see that metrology has at least three parts, each depending on the one before it:

1) A dimension along which to measure, for example, length or mass. Dimensions classify the interactions we (or our instruments) have with other things. Dimensions develop over time after the discovery of new forms of physical interaction. For example, voltage applies to interactions involving electricity, and only arose after electricity was discovered. Similarly, mass and weight were not distinguished before the discovery of universal gravitation and the fundamental laws of motion [Jammer 1997].

2) Definition of units and scales along a dimension are based on something measurable, for example, a meter is the distance traversed by light in vacuum during the time interval of 1/299,792,458 of a second. Units have scales, for

example, the meter includes multiples and submultiples by factors of 10, such as centimeters and kilometers. The definitions of units are used to create primary reference artifacts.

3) Techniques for measuring artifacts in terms of scaled units and uncertainties, such as a line scale interferometer for calibrating graduated scales of length. They are applied to calibrate secondary references, which are disseminated for measurement of in-use artifacts.

NIST is exploring the applicability of physical metrology approaches to information [Gray 1999][Carnahan 1997]. Most of the existing measures address software development, for example, lines of code, and function points. Emerging measures address quality of software, including number of errors, failures, defects, faults, user satisfaction, and user interface. Measures of errors and failures are often based on a size measure such as number of lines of code [Gray 1999].

Information metrology is passing through the same phases as physical metrology. In particular, dimensions are being developed based on the interactions of information systems and their environment. Once a dimension is identified, a qualitative way of describing the ordering of that dimension is developed. For example, when an object is heavier than another object, we describe it qualitatively, but with the identification of weight as a dimension. By quantifying the unit of weight as pounds, for example, we are able to transform a dimension in the observed system to a quantitative formal system where the ordering of the dimension in the observed systems is preserved in the formal system. The problem of finding a quantitative metric that preserves the ordering of the dimension is a difficult problem in information metrology.

Units and measurements in informational metrology have some characteristics not found in physical metrology. For example:

Nonnumeric total orderings: for example, even if the computations of one system can be shown to subsume those of another, it would not be very useful to say that one system does twice the computation of another. Units of computation would be very dependent on the application that uses the system, which determines the usefulness of the additional computations.

Partial orderings: for example, two systems can be overlapping in the kind of information they support. Neither of these systems supports more kinds of information than the other. Partial orderings do not occur in numerical measurements, except when uncertainties cause the possible range of numbers to overlap.

Relative measures: for example, time and space measures depend on factors that vary so much that an exact number would not be widely applicable. For example, these are affected by the amount of data being processed, the software language and compiler being used, and the machine it is running on. One of the primary kinds of performance measures (*complexity*) are functions on the amount of data being processed, rather than a numeric value [Ding-Zhu 2000][Kuperberg 2005].

Boolean measures: for example, a system might provide information that is always true or not, based on information it is supplied. This is an important characteristic even if it is not known which supplied information is false.

The above characteristics are generalizations of the numeric orderings of physical metrology, rather than qualitative approximations of numeric orderings. A qualitative approximation can still be a total order, for example, ranges of temperatures "hot," "warm," and "cold" are totally ordered. When information metrology specifies only some of the orderings, or the orderings do not obey numeric combination rules, it reflects the domain being measured more accurately, rather than approximating it.

Another factor unique to informational metrology is the relation of information in a system to the entities in the real world to which it refers (*fidelity*). Fidelity is a characteristic not found in physical systems, because physical things are not "about" anything else, as information is. This report does not address fidelity. It assumes information and its meaning are already formalized enough to be communicated to and from an automated system. It does not address how a formalization is derived from real-world experience. Information might not correspond to the real world, either because it is inaccurate or incomplete.

Applying the methodology above to information systems, the first step is to identify their dimensions by examining the ways we (or other information systems) interact with them. There are at least two broad categories of interaction with information systems: information supplied to the system, and information retrieved from the system. [1] For example, in a simple database system, we might expect to be able to create records of entities and facts about them, and to retrieve those facts later. A more sophisticated system might support complicated queries on stored facts or calculate aggregate statistics. These categories are obviously too broad to proceed directly to units and measurement. Refinements of them include:

Representation is the kinds of statements that can be supplied to the information system, and the kinds that it can produce.
Computation (Inference) is the difference between supplied and retrieved information. Inference is a kind of computation.
Performance is the time and memory taken for information to be supplied to a system, stored, computed, and returned to the user.
User interface concerns efficiency of communication between the system and humans.
Software interface concerns accurateness and completeness of communication between systems.

---

[1] The construction and maintenance of the system itself involves other interactions, those between system development tools and the system developer. This report focuses on the user viewpoint, since it is addressing capabilities of reasoning systems.

Representation specifies the syntax of statements and restrictions on meaning of the statements, where the meaning is defined in model-theoretic terms (see Section 3.1). One representation is as *expressive* as another if it can encode all the meanings of the other language. For example, higher order logics are more expressive than first order logic, which is more expressive than description logic. Expressiveness gives a partial order on representations, because some representations might encode some of the statements of another representation and not others.

Inference transforms represented information supplied to the system to the represented information provided by the system. Whether an inference procedure provides only true statements derivable from the supplied information is called *soundness*, and whether it provides all true statements is called *completeness*. A sound and complete representation has a sound and complete inference procedure, but does not indicate what the procedure is. Some inference procedures might not achieve soundness and completeness enabled by a representation, while others might. This means soundness and completeness are refinements on the dimension of inference, even when they are applied to representations.

Performance of an inference procedure always consumes resources, such as time and memory. Whether an inference procedure takes finite time to determine the truth status of any statement in a finite amount of time is called *decidability*. [2] A decidable inference procedure is guaranteed to take finite time on all sentences. An undecidable inference procedure will take infinite time on some sentences. A semidecidable inference procedure will take finite time on true sentences, but might take infinite time on false sentences. A decidable representation has a decidable inference procedure, but does not indicate what the procedure is. Some inference procedures might achieve decidability enabled by representation, while others might not. This means decidability is a refinement on the dimension of inference, even if it is applied to representations.

Performance of decidable inference procedures is often refined into time consumed as a function of the number of inputs to the procedure. When the function gives the time without specifying constants that might be added or multiplied to the time, it is called *complexity*. A common kind of complexity is P (polynomial time), where time is bounded by a polynomial function of the number of inputs, assuming execution on a sequential machine with unlimited memory. The complexity NP (nondeterministic polynomial time) bounds the time to a polynomial of the number of inputs on an unlimited parallel machine with unlimited memory. NP-hard problems, some of which are inferencing problems, are those to which any other problem with at best an NP solution procedure can be transformed in polynomial time. When a problem has only NP solution procedures and it is NP-hard it is called NP-complete.

Performance characteristics of an inference procedure can change depending on the particular information reasoned over. An unsound, incomplete procedure can happen to be sound and complete for the limited information in an individual system. Likewise an

---

[2] Truth status includes falsity, so decidable inference procedures can disprove false statements in finite time. Complete inference procedures are guaranteed only to prove all true statements.

undecidable representation can happen to be decidable for a particular set of statements. [3] Complexity is also dependent on the particular information in the system, because it usually assumes the worst case, whereas an application of concern may use information that falls into average or best cases. For example, in constraint satisfaction problems, which are NP-complete in general, the "smoothness" property can lead to algorithms that are polynomial in time [Kumar 2005]. Similarly, in the case of temporal reasoning, restrictions to expressivity of the ontology of time lead to polynomial time algorithms [Krokhin 2003].[4]

More specific criteria can be developed with restrictions on the required inferences. For example, if the knowledge of a subject area is represented in an information system, and the system is presented with specific questions about the subject area, a measure of inference capability might be the percentage of correct answers provided by the system against that of an expert. These types of evaluation provide a measure of confidence of the knowledge base and the reasoning system. This technique could apply to evaluation of user interfaces, which concerns the interaction between the user and the system. Evaluation of the system corresponds to its overall utility measured as improved productivity of the users in their particular context.

A primary purpose of this report is to refine the dimensions of representation, inference, performance, user interface, and software interface precisely enough to support development of units and measurement techniques. For example, the concept of representation must be characterized more precisely before one can be said to be more or less expressive than another. The report does not cover refinements that depend on the particular information reasoned over, or provide criteria that depend on application-specific restrictions. Section 3 focuses on representation, inference, and performance, while Sections 4 and 5 address user and software interfaces, respectively.

3 Reasoning

This section addresses the representation and inference dimensions of information systems (see Section 2). Section 3.1 introduces the basic concepts used in the rest of this section. Section 3.2 covers the various representation languages for expressing information that is reasoned about. Section 3.3 covers the reasoning process itself in terms of specific inference procedures.

---

[3] The information might also be inconsistent or circular, even if the representation is sound, complete, and decidable,

[4] Some argue that the relationship between expressivity and complexity can be addressed by devising algorithms that overcome the complexity barrier through decompositions, such as decomposing first order logic into horn clause statements and a set of constraints [Sowa 2003]. We do not have any proof on the equivalence of this approach to traditional mechanisms for first order logic. However, such decomposition techniques have been used in non-linear optimization problems to reduce the complexity of the problems.

3.1 Introduction to Reasoning

For the purposes of this report, *reasoning* is taken as determining the truth status of statements given the truth status of other statements.[5] A common sense notion of truth is used here. We decide for ourselves what is true in the "world," which can be fictional or mathematical, based on whatever conceptualization we have of it. For example, if we decide there are unicorns in the world that feed on certain fictional grasses, then a reasoner can operate on that based on whatever we decide is true of unicorns. The report does not address philosophical and cognitive issues such as what is real, and how we come to accept it as real.

We also assume reasoners operate on symbols, rather than the actual things in the world, and that there is a flexible link between them (*model theory*). Different symbols may refer to the same thing in the world, and the same symbol may refer to different things in the world under different interpretations of the symbols (an important exception is Herbrand semantics, see Section 3.3.1.2). For example, a reasoner may operate on the symbols "John" and "Fred," which are about the same actual person Jane. Or the symbol "John" may be taken as referring to the person Fred in one case, and Jane in another.

Separating symbol and world is important for reasoning in situations where the entities in the universe of discourse are not completely known, for example, the set of all current terrorists, or where hypotheses are made to link a symbol to an entity in the universe of discourse, for example, linking "leader of London cell" to a known individual. It is also useful in defining criteria for evaluating reasoning procedures, where the world acts as the ultimate verifier of symbol manipulations performed by reasoners.

The rest of this section elaborates on the separation of symbol and world. Section 3.1.1 gives the basic concepts in model theory, and Section 3.1.2 applies them to reasoning.

3.1.1 Model Theory

The relation of symbols and world has at least three aspects:

1) Entities in the world and what is true about them, as we conceive it (*universe of discourse* or *domain*).
2) The symbols operated on by the reasoner and how they are combined (*logical syntax*).
3) The relation between symbols and things in the world (*interpretation*).

Specifically:

---

[5] Truth status is often true or false, but varies by logic.

1) Universe of Discourse (Domain)

   The universe of discourse is the set of objects in the world that will be reasoned about, how they are classified, and links between them. For example, John and Mary are people, and participate in a marriage link. Also called the *domain*.

2) Logical Syntax

   The statements being reasoned over must be written down in some way. This is formalized as *syntax*, which gives a set of *symbols* in the language and rules about how the symbols may be combined. Syntax has some symbols defined by the designer of the reasoning system (*nonlogical symbols*):

   - *Terms* refer to objects in the domain, such as John or Mary. Variables are a kind of term, see quantifiers, below.[6]

   - *Predicates* are applied to terms to make sentences, such as "John is married to Mary." Predicates refer to classifications and relations in the domain, such as Person and Marriage.[7]

   Symbol combination rules specify how to use *logical symbols* to create *sentences* (or *statements*) that have some truth status. Common logical symbols are:

   - The equality operator (=), to tell when two terms are referring to the same entity in the domain.

   - Symbols that combine sentences into new sentences, called *logical connectives*:

     Conjunction ("and", notated as $\wedge$) creates a sentence that is true only if all the combined ones are.
     Disjunction ("or", notated as $\vee$) creates a sentence that is true only if at least one of the combined ones are.
     Negation ("not, notated as ~ in this report) creates a sentence from another that is true only if the negated sentence is false, and vice-versa.
     Implication ("implies", notated as $\Rightarrow$) creates a sentence from two others that is true only if the second sentence is true when the first is. It is the same as a combination of negation and disjunction: p $\Rightarrow$ q is equivalent to ~p $\vee$ q.
     Bidirectional implication ("equivalent", notated as $\Leftrightarrow$) creates a sentence from two others that is true only if the two sentences are

---

[6] Terms can also be functions applied to other terms, see footnote 7.
[7] Model theory includes functions in logical syntax. Functions link things to exactly one other thing, and are used to create terms from other terms. They are taken as a kind of relation for the purposes of this report.

both true or both false. It is the same as a combination of two implications: p ⇔ q is equivalent to p ⇒ q and q ⇒ p.

- *Quantifiers* use *variables* as terms to write a sentence about multiple objects in the domain. The *universal quantifier* ("forall", notated as ∀) requires statements to be true for all values of a variable, such as "all husbands are married." The *existential quantifier* ("exists", notated as ∃) requires statements to be true for at least one value of a variable, such as "all wives have a husband."

A sentence without variables is called a *ground sentence*. Terms without variables are *ground terms*.

Some of the standard vocabulary for parts of sentences is:

- An *atom* is the application of a predicate to terms. These have no logical connectives or quantifiers.

- A *literal* is an atom or a negated atom. Literals without variables are informally called "facts."

- A *clause* is a disjunction of literals. Also called *disjunctive normal form* or *clause normal form*. Since ~p ∨ q is equivalent to p ⇒ q, clauses include implication.

Clauses are the basis of an important form of deductive inference procedure, see Section 3.3.1.1.

All logics addressed in this report use the above symbols, and some add others, see Section 3.2.

3) Interpretation (Semantics)

An *interpretation* is a collection of links from terms/predicates to entities in the universe of discourse giving the "meaning" of the terms and predicates. Two things to note about interpretations:

- The same term or predicate can refer to many entities in the domain, that is, the same term or predicate can appear in multiple interpretations linked to different domain entities.

- Terms and predicates do not need to have the same name as the entities they refer to in the domain, for example, the term "John" may refer to a person named "Fred."

A *variable assignment* is a collection of links between variables and entities in the domain they refer to.

Sentences can be categorized relative to an interpretation:

- An interpretation *satisfies* a sentence if the sentence is true for the entities that the symbols refer to, according to the interpretation. For example, if the sentence is "Fred = John", then an interpretation satisfying the sentence would need to link the symbols "Fred" and "John" to the same entity in the domain.

- A *valid* sentence is satisfied by every interpretation (logical truth). For example, in many logics (p ∨ ~p) is always true, regardless of which terms and predicates the statement p refers to in any domain.

- An *inconsistent* sentence has no satisfying interpretation (logical falsity). For example, in many logics (p ∧ ~p) is always false, regardless of which terms and predicates the statement p refers to in any domain.

An interpretation is a *model* of a sentence if and only if it satisfies a sentence regardless of the way variables are assigned to entities in the domain. All of this vocabulary applies to sets of sentences as well. An interpretation can satisfy a set of sentences, the set can be valid, inconsistent, and have models.

A *representation language* is a logical syntax and at least one interpretation for sentences obeying the syntax.

3.1.2 Inference

With the concepts of Section 3.1.1 we can refine the notion of reasoning to *inference*, which means to change a set of sentences written in some representation language by adding and removing sentences. A *rule of inference* does this by adding or removing sentences based on those already in the set. The rules of inference supported by a representation language or reasoning system is an important way to categorize it.

Since the point of inference is to tell us something about our universe of discourse, we want to use rules that only add sentences that are satisfied by the models we are concerned with, and only remove sentences that are not satisfied (see Section 3.2.8 for logics that use sentence removal). In the strictest approach, inference rules are required to do this for every satisfying model, not just the ones we are concerned with. This is called *logical entailment*. For example, in first order logic, some inference rules that hold for all models are:

The *modus ponens* inference rule generates the sentence q from the sentences ~p ∨ q and p (can also be written p ⇒ q and p). For example, the initial sentences might be "if it is raining, the sidewalks are wet" and "it is raining." The generated sentence is "the sidewalks are wet." The rule assumes that (p ∧ ~p) is always false (*law of non-contradiction*).[8]

The *modus tolens* inference rule generates the sentence ~p from the sentences ~p ∨ q and ~q (can also be written p ⇒ q and ~q). For example, the initial sentences might be "if it is raining, the sidewalks are wet" and "the sidewalks are not wet" and the generated sentence is "it is not raining." The rule assumes that (p ∧ ~p) is always false.

The *universal instantiation* inference rule generates a new sentence by replacing a variable with a term in a universally quantified sentence. For example, from ∀x p(x) ∨ q(x) it can generate p(a) ∨ q(a).[9]

The *existential instantiation* inference rule generates a new sentence replacing a variable with a term in an existentially quantified sentence. For example, from ∃x p(x) ∨ q(x) it can generate p(a) ∨ q(a), where the new term not used in other sentences. It represents the term that the existential says exists (*skolem constants* and *skolem functions*).[9]

An *inference procedure* is a particular way of applying inference rules repeatedly. For example, a breadth first modus ponens procedure applies modus ponens to initial statements in the set, then to ones added in the first round, and so on. The sentences found through an inference procedure are *derived* from the original set. Some important characteristics of inference procedures are:

A *sound* inference procedure will only derive sentences that are logically entailed (are true in all models).

A *complete* inference procedure will derive all logically entailed sentences.

Ideally, we would use inference procedures that are both sound and complete, so we can be sure our sentences correspond exactly to what is true in our universe of discourse, but this is not always possible. Sometimes this is due to the inference procedure, sometimes to the representation language, see Section 3.2. If a representation language supports at least one inference procedure that is sound and complete, the representation is also called sound and complete.

Even if an inference procedure is sound and complete, it may have poor time performance on some kinds of sentences. Two kinds of performance characteristics are:

A set of sentences is *decidable* if the inference procedure will prove whether any sentence is logically entailed or not by applying inference rules a finite number of times.

---

[8] Some rule systems take this as the only inference rule, see Section 3.3.1.2.
[9] The term can be a function of variables that are not already existentially quantified.

> A set of sentences is *undecidable* if the inference procedure might take an infinite number of rule applications to prove whether or not any sentence is logically entailed.
>
> A set of sentences is *semidecidable* if the inference procedure will prove logically entailed sentences by applying inference rules a finite number of times, but might take an infinite number of applications to determine that a sentence is not logically entailed.

If all sentences in a representation language are decidable by at least one inference procedure, the representation language is decidable.

An inference procedure is *monotonic* if adding more sentences to an existing set will not reduce the sentences it derives, otherwise, it is *nonmonotonic*. See Section 3.2.8.

Some kinds of sets of sentences can be identified:

> A set of sentences is a *theory* if logical entailment will add no more sentences.
>
> A theory is *complete* if every sentence, or its negation, is in the theory.
>
> A theory is *finitely axiomitizable* if logical entailment can generate all the sentences of the theory from a finite subset of the sentences in the theory.

3.2 Representation Languages

We will use the following three properties of a reasoning system to characterize its representation language:
1. *Theory language* is used to specify the knowledge bases that are used by the reasoning system.
2. *Query language* is used to specify the queries that are solved by the reasoning system.
3. *Ontology* specifies any set of background axioms that are implicitly used by the reasoning system (that is, the ontology is not explicitly represented as a set of axioms within the theory language of query language). For general-purpose reasoners, this set is empty. However, there exist special purpose reasoners that implicitly use an ontology; for example, temporal constraint solvers implicitly use an ontology of time (see Section 3.3.1.3).

In this section, we present a taxonomy of representation languages, ordered by expressiveness. We will use the following characteristics to specify a *profile* for each language in the taxonomy:
1. monotonicity
2. soundness and completeness
3. complexity
4. model-theoretic properties

Additional terminology used in this section is given in Section 3.2.9.

3.2.1 First-Order Logic

Syntax

The symbols of a first-order language are partitioned into logical and nonlogical symbols. The logical symbols consist of:
- variables
- equality symbol
- connectives (negation, conjunction, disjunction, implication, equivalence)
- universal and existential quantifiers

The nonlogical symbols consist of
- constant symbols
- function symbols
- relation symbols

Semantics

A structure or interpretation M for a first-order language L specifies a nonempty set A, the domain, which is the range of variables that occur in sentences of the language.

For each constant symbol c in the language there is an element x in A.

For each m-place function symbol F in the language, there is an m-place function G that assigns a value in the domain for any sequence of arguments in the domain.

For each n-place predicate symbol P, there is an n-place relation R on the set of n-tuples of elements in A.

An interpretation M is a model for a sentence S if the truth value of S with respect to M is true.

Profile

Soundness: Yes.[10]
Completeness: Yes[11].
Complexity: Semidecidable.[12]

Model-theoretic Properties:

The following theorem characterizes first-order logic:

Lindstrom's Theorem

A logic is equivalent to first-order logic if and only if it has the compactness property and the Lowenheim-Skolem property, see Section 3.2.9.[13]

Compactness is required to prove the completeness of the resolution algorithm in refutation theorem-provers.

---

[10] Chapter 2.5 of [Enderton 1972].
[11] Chapter 2.5 of [Enderton 1972].
[12] Chapters 10, 11, and 12 of [Boolos 1980a].
[13] Theorem 1.1.1 in [Flum 1985].

The Lowenheim-Skolem property has philosophical implications for the adequacy of ontologies to capture commonsense concepts [Putnam 1989] [Lakoff 1987].

3.2.2 Restrictions of First-Order Logic

In this section, we consider languages that are restrictions of first-order logic; all such restricted languages are sound and complete. One language L_1 is a restriction of a language L_2 if all of the well-formed formulae in L_1 are also well-formed formulae in L_2 and all models of L_1 are equivalent to models of L_2.

3.2.2.1 Monadic First-Order Logic

This language is the restriction of first-order logic in which all formulae are monadic.

Profile
Complexity: Decidable.[14]

3.2.2.2 Universal-Existential Sentences

In this class of sentences, all existential quantifiers are in the scope of all universal quantifiers, and no universal quantifier is in the scope of an existential quantifier.

Profile
Complexity: Semidecidable

Model-theoretic Properties
The theorem below characterizes the models of any theory consisting of universal-existential sentences.

A theory is preserved under unions of chains if and only if it has a nonempty set of universal-existential axioms[15]

This theorem is used to characterize the models of an ontology and to prove that an ontology is verified.

3.2.2.3 Universal Sentences

This class of sentences allows only universal quantifiers (no existential quantifiers). Universal sentences that are disjunctions of literals are known as clauses, see Section 3.1.1.

---

[14] Theorem 25.1 of [Boolos 1980].
[15] . Theorem 3.2.3 of [Chang 1973].

Profile
    Complexity: Semidecidable

Model-theoretic Properties
    The theorem below characterizes the models of any theory consisting of universal sentences. It is used to characterize the models of an ontology and to prove that an ontology is verified.

A theory is preserved under submodels if and only if it has a nonempty universal set of axioms [16]

The most important property of universal theories is known as Herbrand's Theorem:

A set of clauses has a model if and only if it has a Herbrand model [17].

By this theorem, for a refutation theorem prover (see Section 3.3.1.1) to show that a set of clauses is unsatisfiable, it is sufficient to show it is unsatisfiable for a Herbrand interpretation.

3.2.2.4 Horn Clauses

A sentence is a Horn clause if and only if it is a universal sentence whose clausal form contains at most one positive literal.

Profile
    Complexity: Semidecidable, although propositional Horn clauses can be decided in polynomial time. [18]

Model-theoretic Properties
    The theorems below characterize the models of any theory consisting of Horn sentences.

Any set of Horn clauses has a unique minimal Herbrand model. [19]

This theorem plays a crucial role in the semantics of logic programs (Section 3.3.1.2).

Every Horn theory is preserved by products. [20]

This theorem is used to characterize the models of an ontology and to prove that an ontology is verified.

---

[16] Theorem 3.2.2 of [Chang 1973].
[17] Proposition 3.2 of [Lloyd 1984]
[18] See [Arvind 1987] and [Dowling 1984].
[19] Theorem 6.2 in [Lloyd84].
[20] Theorem 9.1.5 in [Hodges 1993].

3.2.2.5 Datalog

This restriction consists of Horn sentences that do not contain any function symbols.

Profile
   Complexity: Decidable.

3.2.2.6 Existential Sentences

This class of sentences allows only existential quantifiers (no universal quantifiers).

Profile
   Complexity: Decidable

Model-theoretic Properties
   The theorem below characterizes the models of any theory consisting of existential sentences.

   A theory is preserved under extensions if and only if it has a nonempty existential set of axioms.[21]

3.2.3 Beyond First-Order Logic

In this section, we explore languages that are more expressive than first-order logic.

3.2.3.1 Transitive Closure Logic

This language is the extension of first-order logic that is closed under the transitive closure of all definable relations.

Profile
   Soundness: Yes
   Completeness: No
   Complexity: Undecidable

Model-theoretic Properties
   The following theorem characterizes transitive closure logic:

   Transitive closure logic captures the NLOGSPACE complexity class.[22]

3.2.3.2 Least Fixpoint Logic

This language is the extension of first-order logic that is closed under the least fixed-point of all definable relations.

---

[21] Corollary 6.5.5 of [Hodges 1993].
[22] Theorem 7.5.2 in [Ebbinghaus 1999].

Profile
    Soundness: Yes
    Completeness: No. [23]
    Complexity: Undecidable, since the set of formulas valid in all structures is not recursively enumerable. [24]

Model-theoretic Properties
    The following theorem characterizes least fixpoint logic:

Least fixpoint logic captures the PTIME complexity class. [25]

3.2.3.3 Partial Fixpoint Logic

This language is the extension of first-order logic that is closed under the arbitrary fixed-points of all definable relations.

Profile
    Soundness: Yes
    Completeness: No. [26]
    Complexity: Undecidable, since the set of formulas valid in all structures is not recursively enumerable. [27]

Model-theoretic Properties
    The following theorem characterizes partial fixpoint logic:

Partial fixpoint logic captures the PSPACE complexity class. [28]

3.2.3.4 Second-Order Logic

This lanaguge is the extension of first-order logic in which quantifiers range over the power set of the domain.

Profile
    Soundness: Yes
    Completeness: No. [29]
    Complexity: Undecidable. [30]

3.2.3.5 Monadic Second-Order Logic

This language is the restriction of second-order logic in which second-order quantification is permitted only over unary relation variables.

---

[23] See discussion in [Leivant 1994].
[24] See discussion in [Leivant 1994].
[25] Theorem 7.5.2 in [Ebbinghaus 1999].
[26] See discussion in [Leivant 1994].
[27] See discussion in [Leivant 1994].
[28] Theorem 7.5.2 in [Ebbinghaus 1999].
[29] Corollary 18.2 of [Boolos 1980].
[30] Corollary 18.2 of [Boolos 1980].

Profile
    Soundness: Yes
    Completeness: No.[31]
    Complexity:
Monadic second-order logic without functions is decidable. However, with functions, the set of monadic second-order formulas valid in all structures is not recursively enumerable.[32]

Model-theoretic Properties
    The following two theorems characterize monadic second-order logic:

The restriction of monadic second-order logic in which only existential quantifiers over relations are allowed captures the NPTIME complexity class. [33]

Any language accepted by a finite automaton is definable in monadic second-order logic.[34]

3.2.3.6 Infinitary Logic

This language is the extension of first-order logic that is closed under conjunction of arbitrary (possibly infinite) sets of formulae.

Profile
    Soundness: Yes
    Completeness: Yes. [35]
    Complexity: Undecidable

Model-theoretic Properties
    Infinitary logic is not compact (since an infinite sentence may be satisfiable, yet no finite subformula of the sentence is satisfiable).

3.2.3.7 CycL

CycL is a formal language whose syntax derives from first-order logic. However, in order to support the Cyc Ontology and its concepts related to common sense knowledge, the expressiveness of CycL goes far beyond first order logic. In particular, CycL supports a second-order syntax, in which relations can be arguments for other relations; it is not clear whether or not CycL has a second-order model theory. CycL also includes a class theory with distinguished relations for specifying instances of classes, and generalizations of classes.

CycL contains nonmonotonic notions (see Section 3.2.8) in which assertions are default true or monotonically true. Assertions that are monotonically true are held to be true in

---

[31] See discussion in [Leivant 1994].
[32] See discussion in [Leivant 1994].
[33] Theorem 7.5.2 in [Ebbinghaus 1999].
[34] Theorem 6.2.3 in [Ebbinghaus 1999].
[35] See discussion in keisler71.

every case, that is, for every possible set of bindings to the universally quantified variables (if any) in the assertion, and cannot be overridden. Assertions that are default true, in contrast, are held to be true in most cases, and can be overridden.

Profile
  Soundness: unknown
  Completeness: unknown
  Complexity: unknown

3.2.4 Reified First-Order Logics

In this section, we explore languages whose syntax extends that of first-order logic, but whose model theory is still first-order. Such languages are often referred to as reified logics, since the extension allows predicates and function symbols to be arguments of predicates and function symbols, so that in some sense the relations and functions are "things" in the domain.

3.2.4.1 Common Logic

The central intuition for the syntax of Common Logic is that all entities (individuals, propositions, properties, and relations alike) are first-class logical citizens that jointly constitute a single domain of quantification [ISO 2005a]. Hence, such entities can themselves have properties, stand in relations, and serve as potential objects of reference. The resulting syntactic freedom allows a wide variety of alternative first-order axiomatic styles to co-exist within a common syntactic framework, with their meanings related by axioms, all expressed in a single uniform language.

Although the syntax of Common Logic allows quantification over functions and relations, Common Logic without sequence variables is semantically equivalent to first-order logic.

In a higher-order logic, quantification over a higher type is required to be understood as ranging over all possible entities of that type; for example, quantification in second-order logic is over the set of all possible relations (which is the power set of the domain). The goal of higher-order logic is to express logical truths about the domain of all relations over some basic universe, so a higher-order logic supports the use of comprehension principles that guarantee that relations exist. In contrast, Common Logic allows the universe to contain relations, but imposes no conditions on what relations exist other than that every name has a referent.

A sequence variable in Common Logic stands for an arbitrary sequence of arguments. Since sequence variables are implicitly universally quantified, any expression containing a sequence variable is logically equivalent to the infinite conjunction of all the expressions obtained by replacing the sequence variable by a finite sequence of names. This ability to represent infinite sets of sentences in a finite form means that Common Logic with sequence variables is not compact, and therefore strictly not first-order. Common Logic with sequence variables is equivalent to a restriction of infinitary logic.

Profile
    Soundness: Yes
    Completeness: Yes.[36]
    Complexity: unknown

3.2.5 Description Logics

The basic elements of any description logic are concepts (which denote classes) and roles (which denote binary relations) [Calvenese 2001]. The simplest description logic is AL, whose constructs are shown in Table 1.

Different description logics are characterized by the constructors, which can be used to specify arbitrary concept and role expressions from atomic concepts and roles (see Table 2). The description logic S is equivalent to the extension of ALC in which roles are transitively closed.

Profile
    Soundness: Yes
    Completeness: Yes
    Complexity: See Table 3.[37]

Constructor	Syntax	Semantics
primitive concept	A	$A^I \subseteq \Delta^I$
primitive relation	R	$R^I \subseteq \Delta^I \times \Delta^I$
top	$\top$	Δ^I
bottom	$\bot$	$\emptyset$
conjunction	$C \sqcap D$	$C^I \cap D^I$
universal quantification	$\forall R.C$	$\{x \mid \forall y. R^I(x,y) \supset C^I(y)\}$
existential quantification	$\exists R$	$\{x \mid \exists y. R^I(x,y)\}$

Table 1: Constructors in description logic AL

Constructor	Name	Syntax	Semantics
negation	$\mathcal{C}$	$\neg C$	$\Delta^I \setminus C^I$
existential restriction	$\mathcal{E}$	$\exists R.C$	$\{x \mid \exists y. R^I(x,y) \wedge C^I(y)\}$
disjunction	$\mathcal{U}$	$C \sqcup D$	$C^I \cup D^I$
qualified cardinality	$\mathcal{Q}$	$\geq nR.C$	$\{x \mid \sharp\{y \mid R^I(x,y) \wedge C^I(y)\} \geq n\}$
enumeration	$\mathcal{O}$	$\{a_1, ..., a_n\}$	$\{a_1^I, ..., a_n^I\}$
inverse role	$\mathcal{I}$	R^-	$\{(x,y) \in \Delta^I \times \Delta^I \mid (y,x) \in R^I\}$
role inclusion	$\mathcal{H}$	$R \sqsubseteq S$	$R^I \subseteq S^I$

Table 2: Constructors for the extensions of description logic AL

---

[36] This claim is made in [ISO 2005].
[37] See discussion in [Donini 1996] [Donini 1997].

Description Logic	Complexity
$\mathcal{AL}$	P
$\mathcal{ALU}$	NP
$\mathcal{ALE}$	NP
$\mathcal{ALC}$	PSPACE
$\mathcal{ALCO}$	PSPACE
$\mathcal{SHIQ}$	PSPACE

Table 3: Complexity of description logics

3.2.6 Web Languages

Working groups within the World-Wide Web Consortium (W3C) have designed several languages that can be used for applications that need to understand the content of information beyond the human-readable presentation of content.

3.2.6.1 RDF/S

The Resource Description Framework (RDF) is a representation language intended to be used to express propositions using precise formal vocabularies, particularly those specified using RDF Schema (RDFS), for access and use over the World Wide Web, and is intended to provide a foundation for more advanced representation languages with a similar purpose.

Syntax:

> The underlying structure of any expression in RDF is a collection of triples, each consisting of a subject, an object, and a predicate (also called property) that denotes a relationship. A set of such triples is called an RDF graph, whose nodes are the subjects and objects, and whose directed edges are the relationships.
>
> RDF uses a vocabulary based on Universal Resource Identifiers (URI). A node in an RDF graph may be a URI and properties are URI references. A URI reference or literal used as a node identifies what that node represents. A URI reference used as a predicate identifies a relationship between the things represented by the nodes it connects.
>
> RDF is a logic in which each triple expresses a simple proposition. The assertion of an RDF triple says that some relationship, indicated by the predicate, holds between the things denoted by subject and object of the triple. The assertion of an RDF graph amounts to asserting all the triples in it, so the meaning of an RDF graph is the conjunction of the statements corresponding to all the triples it contains.
>
> A complete specification of the syntax of RDF can be found in [W3C 2004d].

Semantics:

RDF does not impose any logical restrictions on the domains and ranges of properties; in particular, a property may be applied to itself. When classes are introduced in RDFS, they may contain themselves. Such "membership loops" might seem to violate the axiom of foundation, one of the axioms of standard (Zermelo-Fraenkel) set theory, which forbids infinitely descending chains of membership. However, the semantic model given here distinguishes properties and classes considered as objects from their extensions - the sets of object-value pairs that satisfy the property, or things that are "in" the class - thereby allowing the extension of a property or class to contain the property or class itself without violating the axiom of foundation. In particular, this use of a class extension mapping allows classes to contain themselves.

The use of the explicit extension mapping also makes it possible for two properties to have exactly the same values, or two classes to contain the same instances, and still be distinct entities. This means that RDFS classes can be considered to be more than simple sets; they can be thought of as "classifications" or "concepts" which have a robust notion of identity that goes beyond a simple extensional correspondence.

An interpretation specifies for each URI reference, what it is supposed to be a name of; and also, if it is used to indicate a property, what values that property has for each thing in the universe; and if it is used to indicate a datatype, that the datatype defines a mapping between lexical forms and datatype values. This is just enough information to fix the truth-value of any ground triple, and hence any ground RDF graph.

RDF can be thought of as a version of existential binary relational logic in which relations are first-class entities in the universe of quantification.

A complete specification of the semantics for RDF can be found in [W3C 2004c].

Profile
Soundnesss: Yes
Completeness: Yes
Complexity: NP-complete

3.2.6.2 OWL
The Web Ontology Language (OWL) [W3C 2004e] is intended to provide a language that can be used to describe the classes and relations between them that are inherent in Web documents and applications. It can be viewed as an extension of a restricted view of the RDF language with constructs from description logics.

The OWL language provides three increasingly expressive sublanguages [W3C 2004b]:

- OWL-Lite has been defined with the intention of creating a simple language that will satisfy users primarily needing a classification hierarchy and simple constraint features. For example, while it supports cardinality constraints, it only permits cardinality values of 0 or 1.
  Profile
      Soundnesss: Yes
      Completeness: Yes
      Complexity: P

- OWL-DL includes the complete OWL vocabulary, interpreted under a number of simple constraints. Primary among these is type separation. Class identifiers cannot simultaneously be properties or individuals. Similarly, properties cannot be individuals. OWL DL is so named due to its correspondence with description logics.
  Soundnesss: Yes
  Completeness: Yes
  Complexity: PSPACE

- OWL-Full includes the complete OWL vocabulary, interpreted more broadly than in OWL DL, with the freedom provided by RDF. In OWL Full a class can be treated simultaneously as a collection of individuals (the class extension) and as an individual in its own right (the class intension).
  Profile
      Soundnesss: Yes
      Completeness: Yes
      Complexity: PSPACE

3.2.7 Modal Logics

In this section, we explore various modal logics and their relationships to each other. A modal logic is a syntactic and semantic extension of propositional or first-order logic with new operators ($\Box$ and $\Diamond$) on sentences. These operators have various intuitive interpretations:

- $\Box$ is necessity and $\Diamond$ is possibility
- $\Box$ is knowledge
- $\Box$ is provability and $\Diamond$ is consistency

3.2.7.1 Propositional Modal Logics

Propositional modal logics have been widely used to represent the knowledge and belief of intelligent agents, to represent temporal information, to analyze the behavior of distributed systems [Fagin 2003] and to evaluate the correctness of ontologies [Welty 2001][Guarino 2003].

Syntax

The symbols of a propositional modal logic consist of the logical symbols of propositional logic together with the two operators, $\Box$ and $\Diamond$, that satisfy the sentence:

$$\Diamond p \Leftrightarrow \neg\Box\neg p$$

Semantics

In modal logic, a model is a triple <W,R,V>, where
- W is a set of possible worlds,
- R is a binary accessibility relation defined over the set of possible worlds,
- V is a function assigning truth values to formulas at each possible world. In particular, V($\Box p$) = true at some possible world W if and only if V(p) = true at all worlds that are accessible from W.

Different modal logics are characterized by properties of the accessibility relation within the models. Each accessibility relation corresponds to a modal axiom (see Table 4). In epistemic logics, where $\Box$ is interpreted as knowledge, the system KT4 captures the intuition of positive introspection (if an agent knows something, then it knows that it knows) and the system KTE captures negative introspection (if an agent does not know something, then it knows that it does not know). The system G captures the logic of provability [Boolos 1993].

Profile

Soundness: Yes.
Completeness: Yes.[38]
Complexity: Decidable.[39]

Modal System	Axiom	Accessibility Relation
K	$\Box(p \supset q) \supset (\Box p \supset \Box q)$	-
T	$\Box p \supset p$	reflexive
D	$\Box p \supset \Diamond p$	serial
4	$\Box p \supset \Box\Box p$	transitive
E	$\Diamond p \supset \Box\Diamond p$	Euclidean
B	$p \supset \Box\Diamond p$	symmmetric
G	$\Box(\Box p \supset p) \supset \Box p$	well-founded

Table 4: Modal axioms

---

[38] See discussion in [Hughes 1996] and [Hughes 1984].
[39] See discussion in [Vardi 1997].

3.2.7.2 Quantified Modal Logics

This language is the extension of first-order logic with the modal operators [Fitting 1998]. However, there is no consensus on which inference rules and axioms should be used, particularly with respect to the domain of quantification [Cresswell 1991] [Linsky 1994]. There are two major proposals -- the possibilist approach assumes a single domain of quantification that contains all the possible objects, while the actualist approach assumes that domain of quantification changes from world to world, and contains only the objects that actually exist in a given world.

Axiomatically, the possibilist approach assumes the Barcan formula:
$\forall x \Box p \Rightarrow \Box (\forall x) p$
that is, if everything necessarily possesses a certain property, then it is necessarily the case that everything possesses that property. However, this can also lead to counterintuitive inferences with respect to identity [Carnap 1947].

3.2.8 Nonmonotonic Logics

All classical logics in the previous sections are monotonic – any inferences drawn from a theory are preserved by any set of sentences containing the theory. However, many scenarios in commonsense reasoning do not possess this property, for example, when some inferences depend on the failure of other inferences. If the particular statements that failed to be proved are in a set of sentences that contains the theory, the original inferences will not hold. This might be resolved by removing the results of the original inference, see Section 3.1.2, whereupon the set of sentences is no longer monotonically increasing in size. This arises in reasoning with incomplete information, where inference relies on the absence of information. When new information becomes available, previous inferences do not hold anymore. This means that nonmonotonic logics are in general unsound, since inferred conclusions are not always valid (that is, satisfied by all models of the theory). [40] Various nonmonotonic logics have been proposed to formalize this, as described in the following sections.

3.2.8.1 Reiter's Default Logic

Default logic [Reiter 1980] incorporates inference rules of the form
A: B / C
meaning roughly, "If A holds in the present theory T, and B is consistent with this theory, then infer C." The set of formulae which can be derived from the defaults is called an extension. The set of defaults can be viewed as extending the knowledge contained in T, about an incompletely specified world. An extension is then an acceptable set of beliefs one may hold about the world. In general, default extensions are sets of formulae that only partially characterize the theory, that is, an extension need not be complete.

---

[40] A survey of complexity results for nonmonotonic logics is available in [Cadoli 1993].

Reiter's default logic has been implemented in several systems. In particular, Theorist [Poole 1988] implements a default reasoner as a first-order theorem prover that uses a given set of hypotheses to complete a proof if there are insufficient facts.

3.2.8.2 Model Preference Default Logic

[Selman 1988] proposed a new formalism for nonmonotonic reasoning based on a semantic characterization of default inference and preferred models. They define a preference ordering over the space of all possible models; this is represented in default rules of the form

$A \rightarrow B$

which is interpreted as "a model where A holds is preferred over one where B holds." The goal of default inference is to find a most preferred model.

The application of a default rule d of the above form to a model M leads to a model d(M), which is identical to M with the possible exception of the truth assignment to the letter corresponding to the literal B; this letter is assigned a truth value such that B becomes *true*. If the left hand side of the rule is $\emptyset$ (which represents the empty model), then the literal on the right hand side holds by default. For example, consider the default rules (from [Selman and Kautz 1988], where s is an abbreviation for student, a an abbreviation for adult, and e is an abbreviation for employed:

$s \rightarrow a$

$s \rightarrow e$

$s \rightarrow \neg e$

The first rule says that when given two models that assign *true* to student and that differ only in the truth assignment of adult, give preference to the model with adult assigned *true*.

In this example, there are two defaults whose conclusions conflict -- if someone is both a student and an adult, the defaults conclude that the person is both employed and not employed. To resolve this conflict, the notion of specificity is introduced in order to block the application of defaults. A default d of the form

$A \rightarrow B$

is blocked at a model M if and only if there exists a default d of the form

$A \cup B \rightarrow \neg X$

and M does not satisfy A∪B. In the example, if we introduce the rule

$s \wedge a \rightarrow \neg e$

then the application of the rule

$a \rightarrow e$

is blocked because $s \wedge a$ is more specific than a.

3.2.8.3 Circumscription

Circumscription [McCarthy 1980] [McCarthy 1986] is based on the intuition that rather than inferring sentences that are satisfied by all models of a theory, we want to infer sentences that are satisfied by certain preferred models. In particular, circumscription defines the preferred models to be those in which certain predicates have minimal extensions. This definition of minimality is defined by the introduction of a second-order logic formula. Although circumscription is not directly implemented in this way, it has been influential in characterizing the semantics of logic programs (Section 3.3.1.2.4).

3.2.8.4 Autoepistemic Logic

Autoepistemic logic [Moore 1985] uses a modal operator L for knowledge to specify sentences that represent conjectures based on incomplete knowledge. The default from example from section 2.1.1.2 would be written as

$$a \wedge \neg L \neg s \Rightarrow e$$

meaning that if the agent does not know that some adult is a student, then it will conjecture that the adult is employed. The semantics of autoepistemic logic is based on the notion of expansions [Antoniou 1997], which are pieces of knowledge defining "world views" compatible with and based on the given knowledge. Inferences made with autoepistemic logic are based on stable expansions, which are sets of sentences satisfying the following two conditions:

$$\varphi \in E \Rightarrow L\varphi \in E$$
$$\varphi \notin E \Rightarrow \neg L\varphi \in E$$

Although there are no direct implementations of autoepistemic logic, it has been influential in characterizing the semantics of logic programs (see Section 3.3.1.2).

3.2.9 Additional Terminology

This is additional terminology to Section 3.1 for Section 3.2:

- *Axiomatizes*: A set of sentences in a logical language axiomatizes a class of structures if and only if every model of the sentences is in the class of structures and every structure in the class satisfies the set of sentences. A class of structures is axiomatizable in a language L if and only if there exists a set of sentences in L that axiomatizes the class.

- *Captures*: A language L captures a complexity class C if for any class K of ordered structures, we have
    $$K \in C$$
    if and only if K is axiomatizable in L.

- *Compactness*: A logic has the compactness property if an infinite set of sentences in the language has a model if and only if every finite subset of sentences has a model.

- *Completeness (of representation)*: A representation is complete if for any sentence S and theory T, if S is satisfied by all models of T, then S is provable from T by some inference procedure (see Section 3.1.2). Equivalently, a representation is complete if for any sentence S and theory T, if S is consistent with T then there exists a model of T that satisfies S.

- *Decidable*: A language is decidable if there is an algorithm for checking for the validity of all sentences in the language. Also see Section 3.1.2. This is equivalent to saying that the set of valid formulas in the language are recursive.

- *Expressiveness*: A language L_1 is as expressive as a language L_2 if and only if for every sentence S in L_2 there exists a sentence P in L_1 such that the set of models of S is equal to the set of models of P.

- *Herbrand base*: If L is a first-order language, then the Herbrand base is the set of ground atoms that can be formed by using predicates from L with ground terms from the Herbrand universe as arguments. Also see Section 3.3.1.2.2.

- *Herbrand interpretation*: An interpretation in which the domain is the Herbrand universe, constants denote unique elements in the domain, and there exists a subset of the Herbrand base that is the set of all ground atoms that are true with respect to the interpretation. Also see Section 3.3.1.2.2.

- *Herbrand universe*: If L is a first-order language, then the Herbrand universe for L is the set of all ground terms that can be formed out of the constants and function symbols in L. Also see Section 3.3.1.2.2.

- *Interpretation*: An interpretation M of a first-order language L consists of the following:
    - a nonempty set D, called the *domain* of the interpretation;
    - for each constant symbol in L, the assignment of an element in D;
    - for each n-ary function symbol in L, the assignment of a mapping from D^n to D;
    - for each n-ary predicate symbol in L, the assignment of a truth value to sets of elements in D^n.
- Also see Section 3.1.1.

- *Logic*: a representation language, see Section 3.1.1.

- *Lowenheim-Skolem property*: A logic has the Lowenheim-Skolem property if whenever a theory has an infinite model, then it also has a countable model.

- *Minimal Herbrand model*: A Herbrand model M for a theory T is minimal if the set of atoms assigned *true* by M is contained in the set of atoms assigned *true* in all other Herbrand models for T. For Horn theories, the Herbrand model is unique, and is equivalent to the intersection of all Herbrand models for the theory.

- *Model*: An interpretation M is a model for a theory T if the truth value for each sentence in T with respect to M is true. Also see Section 3.1.1.

- *Monadic formula*: A monadic formula is a first-order formula, all of whose non-logical symbols are predicates with only one argument.

- *NLOGSPACE*: The set of decision problems that can be solved by a nondeterministic Turing machine using a logarithmic amount of memory space.

- *Preserved under extensions*: A theory T is preserved under extensions if and only if any extension of a model of T is a model of T.

- *Preserved by products*: A theory T is preserved by products if and only if every direct product of models of T is a model of T.

- *Preserved under submodels*: A theory T is preserved under submodels if and only if any submodel of a model of T is a model of T.

- *Preserved under unions of chains*: A theory T is preserved under unions of chains if and only if the union of any chain of models of T is a model of T.

- *PSPACE*: The set of decision problems that can be solved by a Turing machine using a polynomial amount of memory and unlimited time.

- *PTIME*: The set of decision problems that can be solved by a Turing machine using a polynomial amount of time and unlimited memory.

- *Semidecidable*: A language is semidecidable if there exists a procedure that behaves as follows: Given a sentence in the language as its input, the procedure eventually halts if the formula is valid, and may run forever otherwise. Also see Section 3.1.2. This is equivalent to saying that the set of valid formulas in the language are recursively enumerable.

- *Soundness (of representation)*: A representation is sound if for any sentence S and theory T, if S is provable from T by some inference procedure (see Section 3.1.2), then S is satisfied by all models of T. Also see Section 3.1.2. Equivalently, a representation is sound if for any sentence S and theory T, if there exists a model of T that satisfies S, then S is consistent with T.

- *Transitive closure:* The transitive closure of a binary relation R on a set X is the smallest transitive relation on X that contains R.

3.3 Inference

The common modes of inference most extensively studied in artificial intelligence research are deduction, induction, and abduction. These categories are not so easily distinguished, due to numerous variations in the features of the representation languages used. For example, deduction is often taken as concluding new facts from existing facts and general knowledge, while induction is creating new general knowledge. However, for some representation languages knowledge and facts are treated equally, so a deductive reasoner for that language can conclude new knowledge as well as new facts. Likewise, abduction usually means determining the causes of observable facts based on general knowledge, as in diagnosis, but this is frequently equivalent to deducing unobserved facts from observed facts based on general knowledge, as in medicine.

Besides the traditional forms of reasoning, there are other forms such as analogical reasoning and temporal reasoning that respond to different concerns. In the case of analogical reasoning, the focus is mapping features of one domain to another, while in temporal reasoning the expressiveness of the ontology of time is critical to its usefulness for specific problems. This report treats temporal reasoning as a kind of deduction, since temporal reasoners embody an implicit subset of first order logical deduction.

In this section we elaborate on each of these areas, review the current state of art, and identify general characteristics of these systems. Induction is a special case, because the usable knowledge captured determines the performance of the system, and hence there are several attempts at developing metrics in the literature. In the case of analogical reasoning, we have not included case-based reasoning, since it is restricted to similarity of examples in the same domain. [41] Section 3.3.1 describes deduction, Section 3.3.2 induction, Section 3.3.3 abduction, and Section 3.3.4 analogical reasoning.

3.3.1 Deduction

This section covers deduction as found in some common reasoning systems. Section 3.3.1.1 describes first order deduction. Section 3.3.1.2 covers rule systems, including logic programming. Section 3.3.1.3 addresses temporal reasoning.

3.3.1.1 First order deduction

Resolution is the primary deductive inference procedure for first order deduction. It is used in first order theorem provers [Riazanov 2002][McCune 1997][Stickel 1994] and in limited forms in rule systems, see Section 3.3.1.2. A standard vocabulary developed around resolution that is important to understanding any form of reasoning related to first order logic. This section reviews the resolution procedure and its associated vocabulary.

Resolution is a kind of inference rule that generates the sentence $q \vee r$ from the sentences $\sim p \vee q$ and $p \vee r$. For example, suppose we start with the sentences

> John goes to the store or John goes to work.
> John does not go to the store or John has food in the house.

---

[41] There is extensive literature on case-based reasoning and evaluation of these systems [Gonzalez 1998].

We know John either goes to the store or not, so we can conclude he either has food in the house or goes to work. Resolution removes (*resolves*) the two elements about John going to the store, and combines the statements into one disjunctive statement (*resolvant*): [42]

>John goes to work or he has food in the house.

A special case of the above is when r = true, so resolution generates q from the sentences ~p ∨ q and p, which is the modus ponens inference rule (see Section 3.1.2). For example, starting with

>John does not go to the store or John has food in the house.
>John goes to the store.

we can conclude

>John has food in the house.

Resolution works with other kinds of statements if they can be transformed into conjunction of disjunctions (disjunctive normal form, where the statements are clauses, see Section 3.1.1). For example, resolution supports rule forms, because p ⇒ q is equivalent to ~p ∨ q (see Section 3.1.1), so the above example can be written as

>If John goes to the store, then John has food in the house.
>John goes to the store.

or alternatively, using *contrapositive* inference (~q ⇒ ~p from p ⇒ q), as

>If John does not have food in the house, then John does not go to the store.
>John goes to the store.

All first order logical statements can be transformed to disjunctive normal form, however:
- Universal quantifiers are moved to the outside of each disjunction, and the existential quantifiers replaced with functions or constants that stand in the value and make the statement true (skolem constants or skolem functions, see Section 3.1.1). [43]
- Resolution does not support reasoning with equality, but additional inference rules called *demodulation* and *paramodulation* are commonly added for that [Bachmair 1995][Carson 1967].
- Resolution does not directly support reasoning with numbers and other datatype operators, but many reasoners are extended for this.

---

[42] Resolution is a special case of the *constructive dilemma* inference rule, which infers q ∨ s from (~p ∨ q), (~r ∨ s), and (p ∨ r). When p = ~r, it is resolution.
[43] When resolution looks for elements to eliminate by matching with their negation, it uses a procedure called *unification* to bind values to the variables, or variables to each other.

As the above examples illustrate, resolution replaces a number of other inference rules, and the clause form normalizes a wide variety of formats. These aspects simplify the construction of reasoners and proofs about their properties. In particular, resolution is:
- Sound (all logically entailed statements are derivable, see Section 3.1.2).
- Not complete (it cannot derive the true statement $p \vee \sim p$ from an empty set of statements).
- Refutation complete (guaranteed to find a contradiction if a set of statements is inconsistent).[44]

Because of the last property, resolution-based provers begin with a set of consistent statements, add the negation of a statement to be proved, and then determine if the new set of statements is consistent. If it is inconsistent, the statement to be proved is true, otherwise it is false.

Many restricted kinds of resolution are distinguished according to how statements are selected to be resolved. These usually reduce completeness. For example:

Unit resolution requires at least one of the statements to have no logical connectors. It is refutation complete for Horn clauses, but not for all clauses.

Input resolution requires at least one of the statements to be from the statements the procedure started with. It is refutation complete for Horn clauses, but not for all clauses.

Linear resolution or *ancestry-filtered resolution* generalizes input resolution by requiring one the statements to be from the statements that the procedure started with, or that one of the statements is derived from the other (does not resolve statements that are siblings in the derivation trace). It is refutation complete.

Ordered resolution arranges the disjuncts of each statement in order, and only resolves the first literal. It is refutation complete for Horn clauses, but not for all clauses.

Directed resolution is a kind of ordered resolution that puts the positive disjunct of the statement first (*backward-directed*) or last (*forward-directed*). These are also called forward and backward chaining, see Section 3.3.1.2.1.

More detail about resolution is available at [Genesereth 1987][Robinson 2001].

First order inference procedures can be tested against a well-known benchmark for automated proof of mathematical theorems [TPTP 2005]. It provides converters from a number of prover formats. An annual contest is held around these benchmarks to encourage improvement in first order provers.

---

[44] This includes equality if it is extended with other inference rules for equality.

3.3.1.2 Rule Systems

Rule systems can be loosely described as an intersection of deduction and conventional programming. They are influenced as much by practice as theory, and arguably the most widely recognized form of reasoning. For example, they include logic programming [Clocksin 2003][Gabby 1998][Doets 1994], as well as expert system production rule approaches of artificial intelligence [Ignizio 1990][Payne 1990].

Rule systems employ restricted forms of resolution, often reducing completeness, see Section 3.3.1.1. Prolog, a founding example of a logic programming language [Clocksin 2003], is often described as using a special kind of linear resolution, but some say it is backward-directed, because it can resolve sibling statements (derived statements not derived from each other) [Ringwood 1989]. [45] Production rules, typical of expert systems, use optimized forms of either forward- or backward-directed resolution, or both [Payne 1990].

The purpose of this section is to clarify how rule systems differ from unrestricted first order deduction. Section 3.3.1.2.1 introduces rule systems through their relation to conventional programming. Section 3.3.1.2.2 covers the fundamental assumptions in rule system semantics. Section 3.3.1.2.3 describes the inferencing capabilities of rule systems. Section 3.3.1.2.4 Semantics continues the description of rule system semantics. Section 3.3.1.2.4.5 refers to some existing implementations these semantics.

3.3.1.2.1 Intersection of Inference and Programming

Rule systems are based on an analogy between particular kinds of inference procedures and conventional functional programming [Warren 1977][VanEmden 1976][Colmerauer 1993]. For example, suppose we have a rule as shown in Expression 1.

> if it is raining outside then the sidewalks are wet

Expression 1: Rule example 1

In the analogy to conventional programming, the rule is treated as a "procedure," where one of the statements in the rule, either the `if` or `then` part, identifies it for "invocation".[46] The other statement in the rule specifies how the rule reacts to being invoked, the procedure "body." The rule can be used in at least two kinds of inference [Newell 1957]:
- It becomes known that it is raining out, so it is concluded that the sidewalks are wet ("forward" invocation).
- It is asked if the sidewalks are wet, so it is asked if it is raining out to determine the answer ("backward" invocation).

---

[45] One of the more recent resolution implementation techniques supports communication between sibling queries and results (tabling, see Section 3.3.1.2.4.5).
[46] In logic programming the `if` part is called the *tail* and the `then` part is called the head, and the rules are notated with the head before the tail, for example as "head :- tail" or "head ← tail". This report uses the if-then terminology and notation throughout, for consistency.

In forward invocation, the rule is identified by the `if` part, and the body is the `then` part. Applied to the example in Expression 1, the assertion that it is raining identifies the rule, and the body establishes that the sidewalks are wet. In backward invocation, the rule is identified by the `then` part, and the body is the `if` part. Applied to the example above, the question about sidewalks identifies the rule, and the body asks another question about raining. These two ways of invoking rules generalize from conventional functional programming, where the there is only one way of invoking procedures, and there is exactly one identifier and one body.

Forward and backward invocation are used in inference procedures commonly called *forward chaining* and *backward chaining*. As in conventional programming, where a function can call another, the execution of a rule body can invoke other rules. In the forward invocation mode, the assertions in the body invoke other rules identified by those assertions. For example, the assertion that the sidewalks are wet may invoke another rule with that assertion as the `if` part, and a `then` part concluding the sidewalk is slippery. In the backward invocation mode, the queries in the body invoke other rules that are identified by those queries. For example, the query about whether it is raining, may invoke another rule with that query as the `then` part, and an `if` part that asks more questions to determine the answer. *Logic programming* refers to the rule systems that descended from Prolog, which uses backward chaining.

Rule systems also support variables in invocation, in a different way than conventional programming. For example, the rule in Expression 1 could have variables for exactly where it is raining, and the location of the sidewalk. Whereas conventional programming does not restrict the variables of procedures, the variables in rules must be part of statements that are true or false. [47] This limitation has the benefit of defining procedures around specific goals, and requiring structure to support achievement of goals that can otherwise be hidden in conventional programs [Bock 2000] [Filman 1990].

Rules may have multiple variables, as conventional procedures do. In the example above, the rule may have additional variables for how hard it is raining and how wet the sidewalk is. Rules generalize typical conventional programming practice by always supporting variables that either input or output depending on how they are invoked. For example, the query above about whether the sidewalk is wet may specify the location of the sidewalk to find out how slippery it is, or specify slipperiness to find the location of a sidewalk with that characteristic. [48]

Rule systems generalize conventional programming by supporting multiple rules identified by the same statement. These are all invoked when the statement is queried or asserted. For example, an additional rule concluding wet sidewalks is shown in Expression 2. When the sytem is asked if the sidewalks are wet, two rules will be invoked, one asking if it is raining, another asking if the sprinkler is on. In functional programming, an invocation identifies one procedure. Object-oriented invocations can

---

[47] This is sometimes called *pattern-directed invocation* [Hewitt 1969][Warren 1977][Feldman 1991].
[48] Programming languages sometimes have variables that are both input and output, but they are not usually applied in the rule system style.

potentially identify multiple procedures, but only one is actually invoked. The capability to invoke multiple rules based on one identifier has the benefit of supporting incremental addition of knowledge sources without changing the procedures that use the rules. It also causes many of the difficulties in managing rule systems, because rules may interfere with each other, conclude the same statements redundantly, produce multiple results that must be chosen from and searched through, and so on.[49]

if the sprinklers are on then the sidewalks are wet

Expression 2: Rule example 2

3.3.1.2.2 *Basic Rule System Semantics*

A fundamental difference between the semantics of rule systems and the logics of previous sections is that rule systems take symbols as their own interpretation (*Herbrand semantics*), rather than separating symbol and world as described in Section 3.1.1.[50] Specifically, the universe of discourse is all the term symbols in the rule system, the (*Herbrand universe*), plus all the possible statements formed from predicates on these terms (*Herbrand base*), that is, those that do not include rules or other statements with logical connectives other than negation. For example, in Expression 1, the Herbrand universe is sidewalks and outside, and the Herbrand base is the statements that it is raining outside and the sidewalks are wet. The interpretation ties each nonlogical symbol in the rule system to itself, and assigns true to false to each statement in the Herbrand base (*Herbrand interpretation*). In Expression 1 in Section 3.3.1.2.1, the Herbrand interpretation of "outside" is "outside," of "sidewalk" is "sidewalk," and the possible interpretations of "it is raining outside" and "sidewalk is wet" are true or false for each. This differs from model-theoretic interpretation, which can satisfy or not a satisfy sentence (make it true or false, see Section 3.1.1).[51]

An important effect of Herbrand semantics is that it builds in *closure assumptions*:[52]

> All entities in the universe of discourse correspond to different symbols (*unique name assumption*).
>
> All entities in the universe of discourse are known (*closed domain assumption*).
>
> All true facts about the entities are known (*closed world assumption*).

---

[49] Generalizations of rules that allow queries of the if part to be intermixed with the assertions of the then part are closer to programming languages than rules, but still support identification of multiple procedures for a single invocation [Hewitt 1969][Filman 1990][Bock 2001].
[50] One of the founders of logic program came to a similar conclusion [Kowalski 2002], retracting earlier arguments that classical model theory was not necessary [Kowalski 1995].
[51] Functions can be used to create unlimited Herbrand universes, but rule systems restrict or eliminate functions to prevent this, see footnotes 53, 69, 71.
[52] Research is ongoing to relax closure assumptions in logic programming [Baral 1994][Heymans 2004].

These assumptions are useful when the entities in the universe of discourse are uniquely identified, finite in number and all known, and when facts about these entities are also limited and completely known. For example:

> Unique name: a customer database is maintained so all the customers on it have some unique identifier. The unique name assumption is not effective for reasoning in situations where the entities in the universe of discourse might have aliases, for example, in a database of terrorists. In particular, it does not support reasoning with equality to determine that two terms refer to the same entity in the universe. For example, the symbols "John" and "Fred" cannot be equal, even if it turns out they are aliases for the same person in reality. [53]

> Closed domain: a customer database is maintained so all customers are on it. The closed domain assumption is not effective for reasoning in situations where the entities in the universe are not completely known, for example, in a database of terrorists. In particular, it does not support reasoning with quantifiers to refer to all the entities in the universe. For example, a universal quantification to check that all terrorists have been caught might be true simply because all known terrorists have been caught.

> Closed world: a customer database is maintained so all information needed about the customers is also on the database. The closed world assumption is not effective for reasoning in situations where the facts about entities in the universe are not completely known, for example, in a database of terrorists. In particular, it does not support reasoning with negation. For example, it might be inferred that searching airline passengers on a particular flight is not necessary because none are known to be terrorists. This has been a central concern of rule system development and many are extended to support reasoning with negation, see Section 3.3.1.2.4.3.

Resolution and other first order inference procedures are more suitable for applications in which the closure assumptions do not apply, because completion statements can be added only for those aspects of the problem where closure assumptions are useful.

Herbrand semantics can still be used to verify reasoning procedures in which closure assumptions are valid, because it is defined in a logical way, just as model theory is, concisely and declaratively, making it easier to check that a reasoner produces the required result. [54]

---

[53] Equality operators under Herbrand semantics can equate variables as long as at least one is unbound, but cannot equate constants or terms resulting from functions, for example, see Prolog [Clocksin 2003].
[54] Languages adhering to Herbrand semantics can also be used to define constructs that do not make closure assumptions, but it is very cumbersome. For example, a user can define name and equality predicates by specifying their effect individually on every other predicate in the system.

3.3.1.2.3 Rule System Inference

The rule system inference procedures described in Section 3.3.1.2.1 are forms of deduction that partially overlap first-order deductive inference. They restrict first order inference, as well as provide additional capabilities. Section 3.3.1.2.2 compared rules systems and to reasoning by their general approaches to semantics. This section compares the characteristics of rule system inference procedures to first order reasoning.

Some rule systems restrict the `if` part of rules to conjunctions of facts ("and"), and the `then` part to at most one fact (Horn clauses, see Section 3.2.2.4). [55] However, more complicated rules can be transformed to equivalent ones adhering to these restrictions [Lloyd 1984], for example:[56]

1. Disjunctions in the `if` part of a rule are equivalent to multiple rules with one disjunct in the `if` part each, and the same `then` parts. For example, Expression 1 and Expression 2 are equivalent to a single rule with a disjunction in the `if` part of rule for each cause of wet sidewalks ("if it is raining or the sprinklers are on, then the sidewalks are wet").

2. Conjunctions in the `then` part of a rule are equivalent to multiple rules with one conjunct in the `then` part each, with the same `if` part. For example, if a rule concludes that both sidewalks and grass are wet ("if it is raining then the sidewalks and grass are wet"), then this is equivalent to augmenting Expression 1 with another rule concluding the grass is wet if it is raining.

Any of the facts in the `if` part can identify the rule for forward invocation. The `then` parts of rules are also restricted to facts, but differ on whether the facts are conjoined or disjoined. Production rule systems use conjoined facts in the `then` part. Logic programming uses either a single fact in the `then` part or multiple disjoined facts. Any of the facts in the `then` part can identify the rule for backward invocation.

Related to the above is that rule systems usually do not conclude general statements, such existentials, or disjunctive ("or") statements, including rule forms (however, see Section 3.3.1.2.4 about concluding disjunctions). [57] For example, if we know $p \Rightarrow q$ and $q \Rightarrow r$, we cannot conclude $p \Rightarrow r$. This is common in subsumption reasoners, for example, to infer "all dogs are animals" from "all dogs are mammals" and "all mammals are animals" [Haarslev 2003][Horrocks 1999]. It also applies to simplifying rules based on known facts, for example, if the rule is that windy, weekend days are good kite days, and we

---

[55] The latter restriction is due to the restriction to Horn clauses, which allows at most on positive literal per clause.

[56] Another transformation is existentials in the `if` part that quantify variables not used in the `then` part are equivalent to universals over the entire rule [Lloyd 1984]. All these transformations are tautologies in first-order logic, and are not necessary in the resolution inference procedure.

[57] Many recent techniques relax this restriction, but only some include negative literals (rules) in the `then` part, and none include nested quantifiers, for example to conclude an existential statement, see Section 3.3.1.2.4.

know Sunday is a weekend day, a rule system cannot conclude that windy Sundays are good kite days [Grosof 2003].[58] Resolution, by comparison, can conclude statements that have logical connectives and quantifiers.

Following the analogy to conventional programming, rule systems make the closure assumptions of Herbrand semantics (unique name, closed domain, and closed world, see Section 3.3.1.2.2).[59] These might seem like restrictions on general inference procedures, but are actually additional information that support more conclusions. For example, a business might have a rule:

> if most customers like a product, then the product will continue to be manufactured.

Resolution and other first order reasoners would not be able to support a decision to continue manufacturing a product based on this rule, because they would try to prove there is something inherent in the notion of customer that they will like the product, regardless of who the customers happen to be. The reasoners do not assume they know all the customers. Closure assumptions add the information that only those who are on record as being customers actually are (*completion statements*) [Clark 1978]. With this additional information, first order inference can use the rule above for its intended purpose. However, completion statements must be updated as the customers change, as well as previous deductions based on the completion statements. This is nonmonotonic reasoning, see Section 3.2.8, and outside first order logic. Rule systems relieve users of writing and updating completion statements, but semantically their capabilities are still nonmonotonic.

A common example of the increased reasoning power given by closure assumptions is in deducing the transitive closure of a relation, for example, all the ancestors of a person following the parent relation.[60] This is beyond first-order logic, because it cannot tell which entities are not ancestors, only the ones that are. The first order statement is shown in Expression 3. It says that the ANCESTOR relation is equivalent to the parent relation or to a series of linked parent relations.

$$\forall x,z \ (\ ancestor(x,z) \ \Leftrightarrow \ (\ parent(x,z) \lor \exists y \ (\ parent(x,y) \land ancestor(y,z) \ ) \ ) \ )$$

Expression 3: Transitive closure example

Without closure assumptions a first-order reasoner will not assume that two entities are unrelated just because there are no facts relating them. For example, suppose the facts were:

---

[58] For applications only needed to conclude ground facts from such rules, an implication in the then part, such as $p \Rightarrow (q \Rightarrow r)$, can be transformed $(p \land q) \Rightarrow r$ [Lloyd 1984].

[59] Programs do not need to make these assumptions, but typically do. They will usually take different objects as distinct (by analogy that separate parts of the memory in which they are stored are separate), and take iteration over objects actually recorded for a certain type as sufficient for examining all objects of that type.

[60] Transitive closure is related to transitivity: $\forall x,y,z \ p(x,y) \land p(y,z) \Rightarrow p(x,z)$. Transitivity has no base relation, such as PARENT, the transitive closure relation, such as ANCESTOR, will be transitive.

> parent(John, Joe)   *John is the parent of Joe*
> parent(Mary, John) *Mary is the parent of John*

The reasoner cannot prove or disprove whether LISA is an ancestor of JOE, because it will not commit to whether or not there are some parent relations it does not know about connecting JOE (or ancestors of JOE) to Lisa.

With closure assumptions on facts, a first order reasoner can tell which entities are not ancestors. For example, the following statement closes the PARENT facts above, by requiring that only they are true about parentage:

$$\forall x,y \, (\, parent(x,y) \Leftrightarrow ((x=John \wedge y=Joe) \vee (x=Mary \wedge y=John)) \,)$$

With this additional information, a first order reasoner can disprove that LISA is an ancestor of JOE, because it can disprove the existential in Expression 3 for LISA.

The example above is supported by rule systems without adding completion statements, because they build in the closure assumption. For example, Expression 3 would be:[61]

> if parent(x,y) then ancestor(x,y)
> if parent(x,z) and ancestor(z,y) then ancestor(x,y)

When trying to prove whether LISA is an ancestor of JOHN, a backward chaining inference procedure will first check if they are related by PARENT. If they are, then they are ancestors, and if they are not, the system will use the second rule to look for a series of linked parent relations that begin with JOHN. Closure assumptions are applied in this process when it reaches the edge of the transitive closure, in this example, MARY, where there are no parent facts with MARY as the child. The rule system will assume it knows all the parent facts, find that the second rule cannot be applied, and conclude that MARY is not in the transitive closure.[62]

Many rule systems reflect closure assumptions in a negation operator (*negation as failure*, notated in this report as "not").[63] To conclude that a statement is not true using this operator, it is enough to show that the statement cannot be proved, rather than prove that it is false. It is sometimes called the *fail* operator for this reason [Grosof 1999]. For example, "not guilty" in a court of law means that guilt cannot be proved, rather than that the defendant can be proved to be innocent.

Failure negation is useful in applications where the system knows all the necessary facts, or can accept not knowing some of them. Failure negation is insufficient for applications that need to prove certain statements cannot possibly be true. For example, mission-

---

[61] The statements are broken up in rule systems to prevent the infinite recursion caused by having ANCESTOR in the `if` and `then` parts.
[62] Rule systems often evaluate conjunctions in order, so once they cannot prove PARENT in the second rule, they will abandon the rule entirely, avoiding the infinite recursion of trying to prove ANCESTOR again.
[63] Some rule systems do not support negation of any kind, with the obvious limitation of asserting and concluding only positive literals.

critical and life-affecting applications such as an avionics system that ensures a plane will not stray from its course by more than a specified amount.

When using failure negation, it is important to distinguish it from linguistic antonyms. Adapting an example from [Alferes 2003],

>   if charged(x) and guilty(x) then convicted(x)

cannot be transformed to

>   if charged(x) and not innocent(x) then convicted(x)

because the intention is GUILTY means the individual can be proved to be not innocent, rather than they cannot be proved to be innocent.

Failure negation also does not usually apply to statements or conclusions of a rule, because it would be asserting that the system cannot deduce something, which is determined by the system, rather than the user or rules.[64] For example, with Expression 1 in Section 3.3.1.2.1, we might know that the sidewalks are not wet, but we cannot assert the failure negation of wet sidewalks. This means we cannot conclude it is not raining (*contrapositive* inference). We also cannot write the rule

>   if not(sidewalks are wet) then not(raining)

The most we can conclude is the failure negation of raining, but we could do that whether or not the sidewalks were wet, as long as there is no fact indicating it is raining. See Section 3.3.1.2.4.4 for more about contrapositive inference.

In many rule systems, rules must be defined to avoid cycles in failure negation. An example frequently discussed is shown in Expression 4.

>   if not a then b
>   if not b then a

Expression 4: Failure Negation Cycle Example

To prove B, a backward chaining inference procedure will try to prove A with the second rule. This requires the system to try to prove b with the first rule, causing an infinite recursion. Yet the rules are satisfied by either A or B, or both being true. These rule systems, including Prolog, are limited to programs with no cycles of failure negation (*stratified programs*). See Section 3.3.1.2.4 for more discussion of failure negation.

Some recent rule systems provide an additional negation operator that does not make closure assumptions ("classical" negation, notated as ~ in this report). For example, the expression

---

[64] Some systems support this, see Section 3.3.1.2.4.

> if charged(x) and ~innocent(x) then convicted(x)

is correct because it reflects the intention that GUILTY means that the system can prove an individual is not innocent. See Section 3.3.1.2.4 for more about classical negation.

Closure assumptions are one of the primary aspects of rule systems that place them between conventional programming and logical reasoners. They limit the entities and facts a reasoner needs to consider, which reduces logical quantifiers to conventional iteration over a finite set of entities. In the vocabulary of logic, closure assumptions reduce first order logic to unquantified (*propositional*) logic, by replacing each quantified statement with separate statements for all the known individuals.[65]

Rule system inference includes deductions from newly asserted facts (forward-chaining, see Section 3.3.1.2.1). For example, if we know "if p then q" and "if q then r", and assert "p", a rule system can conclude "r". Resolution, by comparison, can only deduce statements proposed by the user, as backward-chaining does in rule systems. Logic programming is also limited to backward chaining, though there is recent research to integrate forward and backward chaining [Baumgartner 2004].

3.3.1.2.4 Rule System Semantics Continued

The limitations of rule system semantics and inference described in previous sections led to a significant amount of research, mostly in logic programming, after the invention of Prolog. These developments address negation, both failure and classical, and closure assumptions. They also address conclusion of disjunctions (*disjunctive logic programs*). Research is still ongoing, but the community seems to be settling around the stable approach or its extension to unknown truth values in well-founded semantics, or some combination. This section describes these two approaches and compares them with examples. It covers only declarative semantics, which is used to tell if particular facts are implied by a set of rules, rather than constructive semantics, which is used to find out what those facts are. Declarative semantics is helpful in using a rule system, because it characterizes the possible results, rather than how results are found, while constructive semantics supports implementations to produce those results.

The road to stable and well-founded semantics is long, starting with Prolog with failure negation, followed by Clark completion semantics for failure negation, stratified programs, fixed point semantics of various kinds, and perfect models [Clark 1978][Chandra 1985][Fitting 2002][Apt 1987]. These intermediate results are not covered here, except as needed to explain the stable and well-founded approaches.[66] Section 3.3.1.2.4.1 covers the approach common to most logic programming semantics. Section 3.3.1.2.4.2 describes stable semantics, Section 3.3.1.2.4.3 well-founded

---

[65] A special case of this is to use the terms in the statements as the known individuals (Herbrand universe, see Section 3.3.1.2.2). An important result is that this set of statements is inconsistent (has no satisfying interpretation, see Section 3.1.1), then the original set of statements without closure is also inconsistent, and vice-versa (Herbrand's theorem).

[66] Most work on logic programming semantics occurred after the initial frameworks for nonmonotonic logic described in Section 3.2.8, but were found to be equivalent in some cases [Pryzymusinski 1990].

semantics, Section 3.3.1.2.4.4 compares them and gives examples, and Section 3.3.1.2.4.5 refers to some existing implementations.

3.3.1.2.4.1 Logic Programming Semantics in General

Most kinds of declarative logic programming semantics, including stable and well-founded, check whether particular subsets of Herbrand base (the possible ground facts formed from all the predicates in the system applied to all ground terms, [67] see Section 3.3.1.2.2) follow the rules of a particular program, where "follow" is specified by the particular semantic approach. In the stable approach, a model is a subset of the Herbrand base that contains true ground facts, and the well-founded approach adds subsets for false and unknown facts. The subsets being tested are the "candidate models," and those that follow the rules of the program are the "models" of the program, [68] rather than models in the model theoretic sense. However, this use of the word "model" is so prevalent in logic programming that it is adopted in this section.

When there is more than one model (more than one way to assign facts to each subset consistently with the rules in the program), a model or models is chosen as the meaning(s) of the program, which is its "semantics" or "preferred models." In some approaches, a program will have at most one model, which is immediately the semantics of the program. Sometimes there is no model, whereupon the program has no semantics.

The first step in the determining the semantics of a logic program is typically to reduce rules with variables to ground rules on each entity in the system separately. The possible inferences will be the same, because Herbrand semantics limits the entities in the system to the ground terms in the rules (the Herbrand universe, see Section 3.3.1.2.2). Specifically, each rule is transformed to rules created by replacing all variables with all possible combinations of ground terms the logic program, reducing them to propositional logic.[69]

3.3.1.2.4.2 Stable Models and Semantics

Once variables are removed from the rules, see previous section, failure negations are removed using another transformation, producing a *reduct* program. This can be viewed as "executing" the failure negation portion of rules based on the candidate model (the particular subsets of Herbrand base being tested for rule conformance), then simplifying the rules based on those results. [70] The following procedure is for stable models, which are subsets of the Herbrand base for true facts [Lifschitz 2002][Gelfond 2002].

---

[67] See footnote 52.
[68] They are actually a kind of Herbrand interpretation, see Section 3.3.1.2.2 and [Pryzymusinski 1990].
[69] It is assumed the rule system either does not support functions, which could generate new terms, or that these functions only produce ground terms already in the program (also see footnote 53). This is the basis for research relating logic programming and propositional satisfiability [East 2001][Marek 1999][Niemela 1999].
[70] This is a kind of partial evaluation, which transforms functions based on partial knowledge of their inputs [Lloyd 1991].

1) Remove all failure negations in the `if` part of rules that apply to facts not in the candidate model. These negations will be true so the result of `if` part of rule will have the same effect when they are removed (the `if` part is a conjunction of facts, and empty `if` parts are assumed to be true).

2) If any of the rules have failure negations remaining in the `if` part, delete the entire rule. These negations are for facts that are in the candidate model, so will be false, and cause the `if` part to be false. The rule will not be able to conclude the `then` part.

3) Remove all failure negations in the `then` part of rules that apply to facts in the candidate model. These negations will be false, so the `then` part of the rule will have the same effect when they are removed (the `then` part is a disjunction of facts in logic programs).

4) If any of the rules have failure negations remaining in the `then` part, delete the entire rule. These negations are for facts that are not in the candidate model, so will be true, and cause the `then` part of the rule to be true. This means the `if` part can be true or false and still conform to the rule, so the rule has no effect on the model.

Once failure negations are removed from the rules, conformance of the candidate model to the rules can be checked by requiring that:

> At least one fact in the `then` part of a rule is in the candidate model when all the facts in the `if` part are.

The *stable model*s are the candidate models with the smallest set of facts that conform to the rules, also called *answer sets* [Gelfond 2002]. [71] The smallest models are found by eliminating candidate models that are supersets of others, yielding the *minimal* models. If a program has a single stable model, it is a *categorical* or *canonical* program. If it has no stable models, it is *incoherent*. If it has at least one stable model, it is *coherent*.

Various kinds of stable semantics are derived from the models:

> In the original stable semantics, only programs with one stable model are given semantics. The facts in the model are taken as true. Facts not in the model are false [VanGelder 1991].

> In *cautious semantics*, only the facts that are in all stable models are taken as true (intersection of models, the *least model*). Other facts will taken as unknown. This is a kind of *three-valued semantics* and turns out to be equivalent to well-founded semantics in many cases, see Section 3.3.1.2.4.3 [Pryzymusinski 1990].

---

[71] Some answer sets formulations exclude functions in the program, see footnote 69.

In *answer set semantics*, the facts in each model are taken as separately true, with facts not in each model separately false (or unknown if using classical negation, see below) [Gelfond 2002]. Each model is taken as a "solution" to the program. This follows the view that the program defines constraints on variable assignments, and inference is a kind of constraint satisfaction for those variables [Marek 1999][Niemela 1999].

In *full answer set semantics*, all facts in the rule-conforming candidate models are taken as separately true, with facts not in each model as separately false (or unknown if using classical negation, see below). This expands answer set semantics to include the non-minimal candidate models that still follow the rules of the program.

In *brave semantics*, all the facts in all stable models are taken as true (union of models) [Gelfond 2002]. Any other facts are taken as false. This can produce sets that do not conform to the program. For programs with one stable model, cautious and brave semantics are the same.

The stable approach can be applied to rules that include an "approximation" of classical negation, which is assumed to be part of facts (see literals in Section 3.1.1). Classical negation applied to an atom means the atom is taken as false, not just as unproven. The stable approach is extended for classical negation as follows:

The Herbrand base is extended to include classically negated atoms that appear in the program.

Failure negation in rules is applied to literals (atoms or classically negated atoms), never the other way around.

The first step in the procedure removes failure negation of classically negative facts in the `if` part of rules when the classically negative fact does not appear in the candidate model.

The third step in the procedure for removing failure negation removes failure negations of classically negative facts in the `then` part of rules when the classically negative fact appears in the candidate model.

Candidate models that take an atom and its classical negation as both true or both false are considered inconsistent and discarded.

In answer set semantics, facts not in particular answer set are taken as unknown under that answer set. A fact is false if its classical negation appears in the answer set.

Some systems based on the approach above can only prove classical negation for negative literals that appear in the program (see first extension above), specifically those that use two-valued truth status. This is why it only approximates classical negation in first order inference, which can deduce a negative literal even if it does not appear in the initial set of statements. See Section 3.3.1.2.4.4 for more about classical negation.

3.3.1.2.4.3 Well-founded Models and Semantics

The well-founded approach extends stable models by adding subsets of the Herbrand base for false and unknown facts (*three-valued models*). [72] The procedure for removing failure negations in the last section (producing the reduct) also works for well-founded models if "candidate model" is replaced with "true subset." Failure negation gives the same result for facts in the false and unknown subset, because the fact cannot be proved in either case.

Once failure negations are removed from the rules, conformance of the model to the rules can be checked by requiring that the `then` part of each rule be "as true" as the `if` part, where `unknown` is less true than `true` and more true than `false`. Specifically, one of the following holds for each rule, testing in order:

> At least one fact in the `then` part is in the true subset (the `then` part is true). Since the `then` part is disjunctive, one true fact in it makes it true, along with the entire rule.

> Otherwise, if at least one of the facts in the `then` part is in the unknown subset (the rest being in the false subset, making the `then` part is unknown), then at least one of the facts in the `if` part is in the unknown subset and none are in the false subset (the `if` part is unknown),

> Otherwise, all the facts in the `if` part are in the false subset (the `if` part is false).

The above can be formalized using numeric values 0, 1/2, 1 for the truth values `false`, `unknown`, and `true`, and the following formulas for calculating Boolean combinations:

$$\text{truth}(\sim S) = 1 - \text{truth}(S)$$
$$\text{truth}(S \wedge V) = \min(\text{truth}(S), \text{truth}(V))$$
$$\text{truth}(S \vee V) = \max(\text{truth}(S), \text{truth}(V))$$
$$\text{truth}(S \Rightarrow V) = 1 \text{ if truth}(V) >= \text{truth}(S), \text{ otherwise } 0$$
$$\text{truth}(\forall x\ S(x)) = \min(\text{truth}(S(x)\ x \text{ in Herbrand universe})$$
$$\text{truth}(\exists x\ S(x)) = \max(\text{truth}(S(x)\ x \text{ in Herbrand universe})$$

where max of empty set is 0, and min is 1.

---

[72] Using three truth values is a simple form of epistemic approach. It assumes there is an agent that knows or believes some things and not others. The agent believes facts in the true subset of well-founded models, believes the facts in the false subsets are false, and does not believe or disbelieve the facts in the unknown subset.

The *well-founded models* are the smallest (minimal) candidate models that have the most facts in their unknown subsets. Every program without disjunction or classical negation (*normal logic programs*) will have exactly one of these models, so well-founded semantics coincides with well-founded models for these programs. When there are no unknown facts, the model is a *total model*, otherwise it is a *partial model*. Well-founded models can be considered a "three-valued" version of stable models [Pryzymusinski 1990].

3.3.1.2.4.4 Comparison and Examples

Well-founded semantics and cautious stable semantics are equivalent for logic programs without disjunction and classical negation, but possibly with failure negation. They both maximize the number of unknown facts, and consequently minimize the number of true and false facts. The other kinds of stable semantics have no unknown facts. The often-cited example in Expression 4 illustrates this:

    if not a then b
    if not b then a

To find the stable models, we can enumerate the possible candidates, remove the failure negations based on those, and test the candidates against the reduced program:

    { } (no facts true)
Removing failure negation leaves two rules:
    if true then a (a is true)
    if true then b (b is true)
The candidate model does not conform to the reduced program. It is not a stable model.

    { a }
Removing failure negation leaves one rule:
    if true then a
The candidate model conforms to the reduced program. It is a stable model.

    { b }
Removing failure negation leaves one rule:
    if true then b
The candidate model conforms to the reduced program. It is a stable model.

    { a, b }
Removing failure negation leaves no rules.
The candidate model conforms to the reduced program. However, it is a superset of another candidate, so is not minimal. It is not a stable model.

The result is that Expression 4 has two stable models, one containing A and another containing B. The various semantics give these results:

> The program has no stable semantics in the original sense, because there is more than one stable model.
>
> Cautious semantics takes A and B as unknown, because there is no overlap between the stable models.
>
> Well-founded model and semantics is the same as the cautious semantics.
>
> Answer set semantics accepts both models, one with A true or unknown and B false, and the reverse in the other. The full answer set semantics adds the non-minimal model containing A and B, taking these facts as true and none false or unknown.
>
> Brave semantics is the same as full answer set semantics in this example.

Most of the above semantics are an advance over their predecessors in handling cycles in failure negation (non-stratified programs) such as Expression 4. This is important when rules are developed in a distributed environment, where cycles may occur and stratification is not possible, for example, rules developed by multiple sources on the web [Alferes 2003].

Each kind of semantics is suitable for particular kinds of applications:

> The original stable semantics is appropriate for those applications that need to be the extremely conservative in deduction. However, this might be considered overly strict, because it will give no semantics at all if there are any facts that are true in only some cases (more than one stable model). In the example above, this semantics will not even say that A and B are unknown.
>
> Cautious stable semantics and well-founded semantics are useful in applications that need to deduce facts that are true in all possible cases the reasoner can find (intersection of stable models). This gives the "skeptical" result, believing only what is absolutely provable, which might be too restrictive for logic programs with disjunction. In the example above, this semantics will not take A or B as true, even though these would be consistent with the program.
>
> Answer set semantics is for applications that use logic programs to explore the various possible situations that are consistent with the program. Answer sets are the possible solutions, and full answer set semantics extends this to non-minimal solutions. Full answer set semantics is useful for logic programs with disjunction. In the example above, answer set semantics will take A or B as true, or both in the full semantics, which are all the solutions.

Brave semantics is sometimes the same as full answer set semantics, and can be useful in logic programs with disjunction, but can also produce facts that cannot be all true at once. It happens to work well in the example above, giving the same solutions as full answer set semantics.

Brave semantics fails when the union of stable models gives facts that are inconsistent when they are all true at once. For example:

> if a then b
> if c then not b

There are two stable models, one containing A and B, and another containing C. Brave semantics would take all of these facts as true at once, leading to the contradiction that B is both true and unprovable. In some cases it can even take a fact and its classical negation as true at the same time.

The well-founded approach supports contrapositive inference (see Section 3.3.1.2.3), but cautious stable only does if it is two-valued. For example, in Expression 1 if we add the fact

> not(sidewalks are wet)

then well-founded semantics will have both raining and sidewalks wet as false, but cautious stable only will if facts not in the stable model are taken as false. This is not the usual case, especially when it is applied to programs with classical negation [Gelfond 1991]. For example, if we changed the above fact to classical negation:

> ~sidewalks-wet

then (three-valued) cautious stable semantics will take raining as unknown, which does not conform to Expression 1, because unknown cannot imply false. This is an effect of using Herbrand semantics on two-valued models, which cannot have ~ RAINING in the model, because it is not in the program. Well-founded semantics has a false Herbrand subset, so it can assign RAINING to that, as well as SIDEWALKS-WET, which conforms to Expression 1.

Handling contrapositive inference properly under cautious stable semantics requires writing separate rules for the contrapositive, for example:

> if ~sidewalks-wet then ~raining

However, the number of additional rules increases with the number of facts in the `if` part, one rule for each negation of the `if` fact.

Disjunctive programs cause problems for both stable and well-founded approaches. Most of the stable semantics do not find all answers in disjunctive logic programs, and well-founded semantics might not produce a single result as it usually does. In addition, the two semantics are not equivalent in some cases. A simple example is

    p or q (equivalent to the rule "if true then p or q")

This has two stable models, one containing P and another containing Q. The cautious stable semantics will take both P and Q as unknown, in its typical skeptical style. Answer set semantics will take one or the other as true, but not both, following its minimal approach, and imitating exclusive disjunction. Full answer set semantics is equivalent to the classical, inclusive disjunction, taking one or the other or both as true. Well-founded semantics tries to maximize unknown facts, but results in four models, two each for P or Q unknown, combined with the other true or false. It cannot take both P and Q as unknown, which would be a unique, maximally unknown semantics, because it would not conform to the program (true cannot imply unknown). The example also shows a case where cautious stable and well-founded semantics produce different results.

A special case of the above example combines disjunction and classical negation [Gelfond 1991]:

    p or ~p (equivalent to the rule if true then p or ~p)

This says that P is not unknown (*law of excluded middle*). Rules like these are not assumed in three-valued semantics, as they are in resolution and first order provers, and affect results when added to programs. For example,

    if p then q

has Q and P unknown in cautious stable and well-founded semantics. Adding the law of excluded middle for P produces two answer sets, one with ~P and another with P and Q.

The above effects of disjunction can appear even if there are no disjunctive rules. For example,

    if ~a then b
    if ~b then a
    if a then c
    if b then c

will not imply C is true under cautious stable and well-founded semantics, because A and B will be taken as unknown (compare to Expression 4). Resolution and other first order provers will infer C, because they will infer that A and B cannot both be false. Disjunctive programs are topics of ongoing research [Minker 2002][Wang 2005].

3.3.1.2.4.5 Implementations

The stable and well-founded approaches have some high-performance academic implementations. Many are for answer set semantics [Niemela 2000][Lierler1 2003][Lin 2004][Giunchiglia 2005]. One supports disjunctive programs [Eiter 2003]. Some academic benchmarks and challenge problems are also available [Hietalahti 2000][Trento 2005]. The main implementation of well-founded semantics also supports answer set semantics and other logics [Rao 1997][SourceForge 2005]. It is based on a very flexible and efficient optimization that records intermediate results for reuse, and notifies intermediate queries of new results (*tabling*) [Chen 1996]. Dependencies are recorded between results to detect cycles, including negation failure cycles. It is an extension of the de facto standard abstract machine for Prolog [Sagonas 1998][Warren 1983], and has a software interface for answer set programming [Castro 2005].

3.3.1.3 Temporal Reasoning

The primary issue in temporal logic in comparison to classical logic is that different statement may have different truth values at different times. Temporal logic allows for extensions to propositional calculus to include statements of the form; A is true at some future time, A was true in some past time, A will be true at all future times and A has always been true in the past.

3.3.1.3.1 Types of Temporal Reasoning System

There are two types of temporal reasoning systems. The types are distinguished based on the underlying ontology of time used by the systems. They treat time as points in the real line or treat time as intervals in the real line. Point-based systems handle point relationships such as (<, =, <=, not=). Interval-based systems use time-intervals as their ontological primitive and use some disjunctive information about time intervals such as the interval is disjoint with, overlaps with, or meets with. The other differences are related to whether they handle only quantitative information or only qualitative information or both. One of the systems handles both, and uses point-based relationship for quantitative constraints, and Allen's interval-based model with qualitative constraints including the disjoint constraint incorporating transitive closure on time intervals.

Two exemplar models of temporal reasoning are:

> *McDermott's model:* In this logic the relation of temporal precedence is transitive, left linear, infinite in both directions, dense and continuous [McDermott 1982]. It employs many-sorted first order predicate logic, with variables permitted to be over the basic ontology of time, states, fact, and events. There are critiques of incompleteness of McDermott's logic. It uses a point-based relationship of time.

> *Allen's model:* Incorporates temporal reasoning and formalizes types of knowledge required to reason about actions and events [Allen 1983]. It uses

many-sorted predicate calculus, with the variables that have a range of properties, time intervals, events, and other primitives.

Temporal relationships between intervals are:
   During (i1,i2)
   Before (i1,i2)
   Overlap (i1,i2)
   Meets (i1,i2)
   Equal (i1,i2)

These relationhips are governed by a set of temporal axioms. There is a concept of point interval denoted by *Point(i1)* and a point interval is not allowed to overlap with other intervals or contain other intervals. There are other primitives such as *Hold (p,i)* where the property holds in interval i. It also includes logical connectors (and, or, not) and quantifiers (universal, existential), see Section 3.1.1. An event is introduced as a new primitive, *Change-pos, with* 4 arguments of objects, the source, goal location, the event itself. The primitive *Occur* has two arguments, event and time interval, which lead to assertion of necessary and sufficient conditions for event occurrence.

Study of Allen' algebra has been an active area of research. In a theoretical effort by [Krokhin 2003], they identified a set of tractable subalgebras of Allens' algebra for describing the ontology of time. In this work, they shown that there are 18 tractable (polynomial time) algebras for temporal reasoning. For details on the 18 subalgebras that are tractable, see [Krokhin 2003]. This work is a generalization of a number of earlier work on identifying tractable and approximate algorithms to address Allen Interval algebra which has been shown to be NP-complete [Vilian 1989][Golumb 1993].

In a separate effort by [Kumar 2005], a special property of constraint satisfaction algorithms called "smoothness" was identified. He shows that satisfaction problems that satisfy the smoothness constraint are solvable in polynomial time. Evaluation of complexity of temporal reasoning systems can be done in terms of Krokhin's subalgebras or the use of the smoothness property. There is no work connecting these two results where a tractable subalgebra has been shown to satisfy smoothness property of a CSP. Both the above citations are very recent, so the lack of the connection is not surprising.

There are other models for temporal reasoning described in [Turner 1996]. We will not elaborate those models as the above two models cover most of the literature in temporal reasoning and are the most commonly studied and used. We provide a summary of evaluations of temporal system presented by [Yampoortam 1993].

3.3.1.3.2 *Evaluation of temporal systems*

Tests on temporal reasoning systems were done using data sets on train schedules. The data sets were generated using a data set generator that incorporated all the temporal constraints needed to schedule trains. The approach was to suppress information such as disjoint constraints to allow for testing of systems that did not handle these constraints.

Table 5 gives a characterization of the temporal systems evaluated in [Yampoortam 1993].

System	Quantitative (point) constraints	Qualitative (interval) constraints	Disjoint Constraint	Transitive closure	Constraint Satisfaction
TimeLogic	x	x			
MATS	x	x	x	x	x
Tachyon	x	x[73]			x
Timegraph II	x	x		x	
TMM	x		x		x

Table 5: Characteristics of temporal system evaluated [Yampoortam 1993]

As the assumptions of the systems were different, a direct comparison was not possible so the performance times considered were:

a) Loading data from files into the right data structures.
b) Executing intermediate calculation steps.
c) Retrieving and/or computing the answers to queries.

As it is not possible in some systems to measure a) and b) they were aggregated for the calculation of "assertion" time and c) was treated as "query" time. The results are presented in [Yampoortam 1993]. Among the two classes of systems evaluated, one that uses incompletely connected graphs perform better in "assertion time" and the other based on fully connected graphs perform with constant query time. It appears that if "assertion time" is not important then systems that use fully connected graphs are better and can use constraint satisfaction methods. It appears that Interval-based relations are better suited for planning and natural language analysis. The choice of the system is dependent on the problem solved, the needed expressivity, and the type of reasoning provided and needed for the problem at hand.

3.3.2 Induction

Induction has been the mainstay of learning in artificial intelligence. In general, induction is defined as the process of finding generic description of concepts for differentiation of a given set of examples.

From the early efforts of Simon and Lea [Simon 1974], Dietterich and Michalski [Diettterich 1981], Quinlan [Quinlan 1986], developing algorithms for classification of a given set of objects with attributes, induction has taken on several forms. These include statistical learning, neural networks, instance based learning, Bayesian learning, genetic algorithms decision rule induction, induction logic programming, analytical learning, reinforcement learning, supervised and unsupervised learning, clustering, and vector machines. Developing metrics for the entire field of induction and machine learning

---

[73] Translated qualitative constraints into quantitative uses epsilon and infinity.

would be a very extensive task and is not possible within the scope of this report. Readers can refer to the review of machine learning by [Flach 2001] [74]. Machine learning is unique in that it does not start with a structured knowledge base in a given representation language and the inference mechanism uses the knowledge to answer questions that are not explicitly encoded in the knowledge base. The validity of a machine learning (induction) algorithm is its ability to generalize from examples and use the generalizations to classify unseen examples. The focus of the performance of a machine learning system is directed at its ability to learn the underlying generalization from examples, with as minimal a set of examples as possible. This changes the scope of performance evaluation of learning systems to specific qualities of the learnt knowledge and the performance of the system. There have been extensive but localized efforts in the evaluation of machine learning systems.

3.3.2.1 Definitions of Induction

The following are some of the machine learning approaches to induction. This is not exhaustive.

- *Clustering:* Clustering is a learning method where the examples are clustered using the attribute value of the entities in the example. Several early machine learning algorithms were clustering algorithms. In the clustering method, the tree that is generated has each node and its sub-nodes as clusters of examples.

- *Rule-induction* (decision tree induction): Rule induction is the generation of decision rules from a set of data. The process of rule induction is often described as a search over a hypothesis space of possible rules for a decision rule that satisfies some criteria. In the rule induction model, a decision tree is generated where the nodes are tests or design rules. At each node, when a test or decision rule is applied, its lead to the choice of a node and its associated test for further refinement of the classification. The non-leaf nodes contain conditions for classification, and the leaf nodes are classes. There are two variations of decision tree induction learning algorithms:

    - *Supervised learning:* Supervised learning is a process in which some initial classification is provided along with the data for the induction process to derive the rules or the classification hierarchy. This approach uses the initial classification of an example by computing distance measures [Quinlan 1993].

    - *Unsupervised learning*: Unsupervised learning is when only the base data is provided to the learning algorithm. No apriori classification of information is used in the distance measure.

---

[74] For an earlier classification of induction see [Cohen 1982].

- *Bayesian Learning:* Bayesian learning is a graphical model that uses Bayesian nets as representation for the classification process. Bayesian inference is integrated with learning in these models. These are probabilistic models that identify joint probability distributions over sets of random variables. They create a probabilistic inference graph by calculating the conditional distributions of the dependent variables given observed variables and minimizing the effect of the non-observed variables. The problem of creating these inference graphs is intractable in general, leading to incorporation of independence assumptions among a subset of variable resulting in directed acyclic graphs. Representations used by researchers are diverse in the encoding of independence assumptions. These are called by different names: Belief networks, Bayesian networks, causal networks, and influence diagrams. There are new extensions to Bayesian networks such as Hierarchical Bayesian Networks that use knowledge about the structure of the data to introduce bias to improve inferencing and learning methods [Gyftodimos 2004].

- *Induction Logic Programming (ILP):* Induction logic programs work on the same principle of generalization from examples. The induction process uses Horn clauses where a predicate is constructed from a set of ground facts. There are two types of induction process followed in ILP. Top-down induction consists of generating hypotheses and testing against examples to identify the predicates. Examples of such systems are Shapiro's Model inference system [Shapiro 1981] and Quinlan's FOIL system [Quinlan 1991]. In this approach, they choose short clauses and add literals until it becomes general. Bottom-up induction starts with the bottom clause, constructed from the ground clauses, and generalizes the clause by throwing out as many literals a possible [Flach 1998].

The advent of data mining from data bases led to efforts such as [Kamber 1997] that address efficiency and scaling of decision tree induction by combing with attribute-oriented induction methods with relevance analysis. These efforts combine different methods of induction to achieve better learning performance.

3.3.2.2 Measurement for Machine Learning Systems

We have summarized two approaches found in the literature for the measurement of performance of machine learning systems. In general, machine learning tools learn from a set of data called "learning set" that are initial set of examples. What is learnt is tested against a set of data termed "Test set."

3.3.2.2.1 Evaluating Machine learning systems

In an overview of applying metrics to machine learning tools, [Alonso 1994] provides a methodology for evaluating the machine learning tools. In this model, a benchmark specification has three components:

a) *Quality characteristics*: Dimensions of the systems measured.

b) *Benchmark suite*: The set of examples or test cases that are used in the benchmarking of the system.
c) *Analysis function*: The authors make the case for an effective measurement system to be implemented the rationale for choosing quality, benchmark suite and analysis functions have to be justified.

Quality Characteristics: [Alonso 1994] identified dimensions for characterizing machine learning systems that were chosen by experts in knowledge engineering and machine learning. The following dimensions were chosen based on weighting them and discarding those that were not deemed relevant by the experts. The dimensions are:

- *Accuracy*: Refers to classification accuracy of the acquired knowledge.
- *Structured Attributes*: Attributes whose values are not flat and form a structure. These correspond to initial classification given to the learning system as in the case of supervised learning.
- *Ordinal Attributes*: Attributes in the learning set whose values can be ordered. It is considered ordinal when the values can be ordered (for example, white, grey, and black) and does not include numeric attributes.
- *Numeric attributes*: A kind of ordinal attribute where relations such as greater-than, less- than are applicable. Traditional dimensions in the measurement of qualities in the physical world such temperature, distance etc.
- *Nominal Attributes*: This attribute identifies the basis for the algorithm and has no ordering.
- *Cost*: Cost associated with test needed to establish an attribute value. Is the cost of obtaining an attribute value used in solving the problem so as to choose the most economically determined information and defer the use of high cost attribute.
- *Noisy and Incomplete Data*: Level of nosiness or incompleteness in the data.
- *Incremental Learning*: The ability to incorporate new examples without recomputing across all the existing examples to derive the set of generalizations.

Table 6 identifies a subset of dimensions that were used in the characterization of the systems evaluated in this paper.

Benchmark Suite: The choice of tasks for each of the dimensions of the system being measured. The tasks are selected carefully. They are based on making sure that the learning set and the test set have examples that contain attributes that target the specific dimension. The learning will be identified with examples that are sufficient for the humans to learn the task and the test suite will be identified with examples that humans find it difficult solve. The choice of learning suite and test suites should be kept as small as possible.

Analysis functions: The authors identify the analysis functions as those that take raw measurements and transform them into meaningful measurements. The approach taken is that every dimension identified as a measurement, an analysis function translates them into a scale for comparison of the systems.

	ASSISTANT 86	ALEXIS II	AQ15	ID*
Ordinal attribute	YES	YES	NO	NO
Numerical attribute	YES	YES	NO	NO
Structured attribute	NO	YES	NO	NO
Cost	NO	YES	YES	YES

Table 6: Dimensions for characterizing the sample set of systems [Alonso 1994]

The above dimensions were tested on some very well known machine learning systems including Assistant 86 [Cestnik 1987], Alexis [Nunez 1991], AQ15 [Alvaro 1990], and ID* [Fernández 1990] family of systems. The sample analysis functions for the different dimensions as provided in [Alonso 1994] are shown in Table 7. Table 8 provides a sample of results from the evaluation in [Alonso 1994].

Ordinal attributes	$f(p) = \max(0, \frac{p - 0.2}{0.8})$
Numerical attributes	$f(x, y) = \begin{cases} 0 & x < 0.5 \\ \frac{(x + 2y)}{3} & x \geq 0.5 \end{cases}$
Structured attributes	$f(x, y) = \sqrt{f_1(x) * f_2(y)}$ $f_1(x) = \max(\frac{x - 0.5}{0.5}, 0)$ $f_2(x) = \max(\frac{x - 0.33}{0.67}, 0)$
Cost	$f(c, p) = \sqrt[3]{g(c) * h(p)}$ $g(c) = \sqrt{1 - c^2}$ $h(p) = e^{(1.5 - (p + 1/2p^2))} + 0.05 \sin^2(\prod(1-p)^{1.8})$

Table 7: Analysis Functions for the dimensions [Alonso 1994]

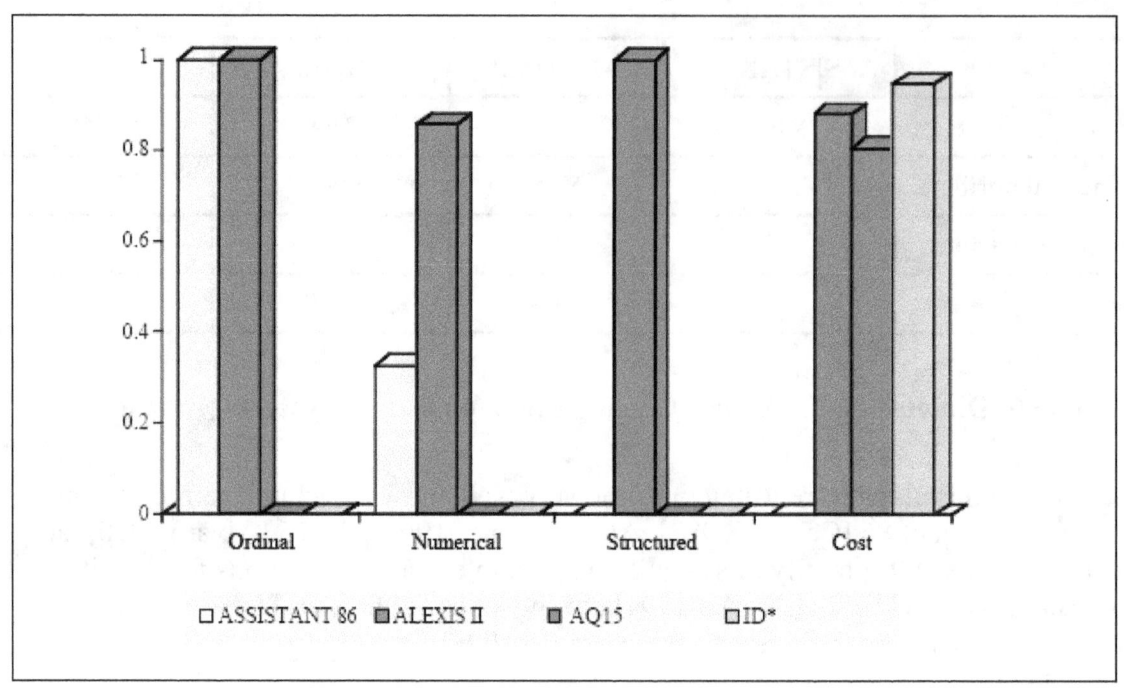

Table 8: Comparative strengths and weaknesses of tools [Alonso 1994]

3.3.2.2.2 Rule Quality Measures for Rule Induction Systems

This section is based on [An 2001], where they have summarized all the rule quality measure in the literature for the evaluation of rule induction systems. The following is a summary of the measures presented in their paper.

Measures for rule quality from rule induction are based on the relationships between a decision rule R and a class C. The relationships can be depicted in a contingency matrix of 2 x 2 of absolute frequencies and relative frequencies for each cross-tabulation. In Table 9 and Table 10, n_{rc} is the number of training examples covered by rule R and belonging to class C, $n_{rc'}$ is the number of examples covered by rule R not belonging to class C, $n_{r'c}$ is the number of examples not covered by rule R belonging to class C and $n_{r'c'}$ is the number of examples not covered by R and not belonging to class C. $n_r, n_{r'}, n_c, n_{c'}$ are marginal totals. For example n_r is the number of examples covered by rule R. Table reflects relative frequencies where, for example $f_c = n_c/N$, and so on.

Class		C Not	Class C	
Covered by rule R		n_{rc}	$n_{rc'}$	n_r
Not covered by R		$n_{r'c}$	$n_{r'c'}$	$n_{r'}$
		n_c	$n_{c'}$ N	

Table 9: Contingency Table with Absolute Frequencies

Class	C	Not Class C	
Covered by rule R	f_{rc}	$f_{rc'}$	f_r
Not covered by rule R	$f_{r'c}$	$f_{r'c'}$	$f_{r'}$
	f_c	$f_{c'}$	1

Table 10: Contingency Table with relative frequencies

3.3.2.2.2.1 Empirical Formulas

Besides the above measurement of rule quality there are several other empirical formulas. These are called empirical formulas, since they are ad hoc rules that have been devised by researchers without necessarily any regards to statistical or information theoretic concepts. They were derived out of the researchers' intuition. Based on the above contingency tables we will define two measures: consistency (also known as accuracy) and coverage. Consistency is defined as $acc(R) = n_{rc}/n_r$ and coverage is defined as $cover(R) = n_{rc}/n_c$. Two formulas that use these two functions:

Weighted sum of accuracy and coverage: It is defined as
$$QW = w1 \times acc(R) + w2 \times cover(R)$$
Specific systems choose specific weights and the sum of the weights are equal to 1. The weights can be used biased towards accuracy or coverage.

Product of Accuracy and Coverage: It is defined as
$$QP = acc(R) \times f(cover(R))$$
where f is an increasing function. The function f is chosen after enough experimentation to ensure that coverage has low impact on the rule quality.

3.3.2.2.2.2 Measures of Association

There are two types of measures of association based on the relationships between the classification for the columns and rows in Table 1. The measures are:

Pearson Chi-square Statistic: This is based on the assumption that the classification for the columns is independent of the rows. Then the frequencies in the cells in the contingency table are proportional to the marginal totals. The chi-square value is given by the traditional Chi-square expression: $SUM((n_o - n_e)^2 / n_e)$, where n_o is the observed absolute frequency of examples in a cell and n_e is the expected absolute frequency of examples in a cell. $(n_o - n_e)^2 / n_e$ is computed for each cell in the contingency table and added to yield the Chi-square value. $n_o = n_{rc}$ and $n_e = n_r n_c / N$. These expressions can be substituted in the Chi-square expression to arrive at one that is completely based on absolute frequencies.

G-square likelihood ratio statistic: The G-square likelihood ration measures the distance between the observed frequency distribution among classes satisfying rule R and the expected frequency distributed of the same number of examples based on the rule R selecting samples randomly. The value of the g-square ratio can be computed using the absolute frequencies in the contingency table. When the G-square ratio is low, then the possibility that there is an association between the observed and expected frequency distribution is also low.

3.3.2.2.2.3 Measures of Information

The measure of information is a statistical measure for rule quality. Given a class C and a rule R, the measure is given by the sum of the amount of information required to correctly classify an instance into class C with a prior probability and the amount of information required to correctly classify an example into class C with a posterior probability of C given R. The measure, called *information score,* is computable using absolute frequencies to estimate probabilities as follows:

$$Qis = - \log(n_c/N) + \log(n_{rc}/n_r)$$

3.3.2.2.2.4 Measures of Logical Sufficiency

The logical sufficiency measures standard likelihood ratio statistic. Given the rule R and a class C, the degree of logical sufficiency of R with respect to C is defined in terms of absolute frequencies can be expressed as

$$Qls = (n_{rc}/n_c) / (n_{rc'}/n_{c'})$$

The larger the value of Qls the applicability of rule R is useful for establishing the membership of an example in C and when Qls tends to infinity, rule R is sufficient to establish the membership of an example in class C logically.

3.3.2.2.2.5 Measures of Discrimination

The measure of discrimination is a statistical rule quality measure. It is a measure of the extent to which R when considered as a query can discriminate between negative and positive examples of the class C. The measure in terms of absolute frequencies is

$$Qmd = \log((n_{rc}/n_{r'c})/(n_{rc'}/n_{r'c'}))$$

The formula represents the log of the ratio between the positive and negative odds of discrimination.

3.3.3 Abduction

Abduction was first used in artificial intelligence for medical diagnosis (Pople 1973). Since then there have been many activities in this area as well as debates on the scope and formalizations of abduction. The models of abduction identified in the literature arcan be categorized by the definition of abduction used by these models. There are three types of abduction: creative (selective), best explanation, and coherence models.

3.3.3.1 Best explanation Model of Abduction

The best explanation model of abduction has been advocated by Josephson [Bylander 1990][Konolige 1991]. These models of abduction are formal models that derive the best explanation of the hypothesis by deduction. The two formal models are different from each other in significant ways. The Josephson model is data-driven best explanation model. In this model, the task is to identify the best explanation for the data provided to the abduction system. The Konolige model is different in that it uses the notion of logical consequences – "hypothesis (causes) explain data (effects)." In the Konolige model, the idea is to deduce the effects from the hypothesis and the domain theory.

3.3.3.1.1 Josephson best explanation model

The Josephson model of abduction is as defined below [Bylander 1990]:
Definition 1. An abduction problem is a tuple {D, H, m, p} where
1. D is the set of all data to be explained
2. H is the finite set of all the individual hypotheses
3. m is the map from all subsets of H to the subsets of D
4. p is the map from subsets of H to a partially ordered set representing the plausibility of various hypotheses

A set of hypotheses h is an explanation if it is complete and parsimonious. This implies that $e(h)$ = D and there exists no proper subset of h that explains all the data that h does.

3.3.3.1.2 Konolige Model for Abduction

There are two definitions in the Konolige model [Konolige 1992].

> Definition 1.1: Let L be a first order language. A causal theory is defined as a tuple {C, E, D} where:
> C is the set of causes represented by sentences in L
> E is the set of effects represented by sentences in L
> D is the set of sentences in L that constitute the domain theory.

> Definition 1.2: Let {C,E,D} be a simple causal theory. An explanation of a set of observation O that is a subset of E is the set A that is subset of C such that:
> A is consistent with D. That is, any hypothesis explaining the effects is consistent with the domain theory.
> D contains the logical consequences; "hypotheses (causes) explain data (effects)" for a give A and O.
> A is subset minimal over sets satisfying the first two conditions i.e. there exist no proper set of A that implies O and is consistent with D.

Both models of abduction decribed above assemble the best explanation and use hypotheses (causes) from data (effects) as the basis for generating the explanation. They subscribe to the condition of parsimony of the hypothesis while the Konolige model requires consistency with the domain theory and does not have condition of explaining all the data while Bylander's model requires that all data be explained. In spite of the differences between these models, they have the same order of complexity measure - NP-complete [Bylander 1990][Thagard 1997].

The Bylander model defines additional conditions that restrict the model of data accommodated by their model of abduction. These proposed restrictions are used to improve the measure of computational complexity. They are

> *Independent abduction problems:* They are a class of problems where a composite hypothesis explains the datum if and only if one of its elements explains the datum. That is, a composite hypothesis consists of several independent hypotheses that explain some datum explained by the composite hypothesis. The model allows for the identification of all the sets of composite hypotheses such that each of the elements in a set explains the datum. As there are an exponential number of explanations, they propose an algorithm that has $O(nC_e+n^2)$ complexity for finding an explanation. The emphasis here is not to find all explanations.

> *Monotonic abduction problems:* In this restriction, a composite hypothesis can explain additional datum that are not explained by any of its elements. This situation arises when two hypotheses have an additive effect. In other words, a particular datum can be explained by the composite hypothesis but not by its elements individually. For example, if there is high blood pressure in the patient and two diagnostic hypotheses indicate that there should be some nominal increase in blood pressure but together they can explain the high blood pressure while they cannot explain the same individually. The composite hypothesis does not lose any data explained by any of the hypotheses.

> In the case of monotonic abduction problems, the task of finding an explanation is $O(nC_e+n^2)$ while the task of finding an additional explanation is NP-complete.

> *Incompatibility abduction problems:* This class of abduction problems negates the assumption that any collection of individual hypotheses is acceptable. However, in many cases, the composite hypothesis can have contradictory hypothesis in its collection but a subset that excludes the offending hypothesis is acceptable. The model of abduction {D,H,m,p} is extended to include N that is a set of two tuples with the incompatible pair of hypotheses such that the model is {D,H,m,p, N}.

> In this case both finding an explanation and finding the best explanation is NP-hard. Further, incompatibility abduction problems can be reduced to Reiter's theory of diagnosis and to the theory of belief revision. Both Reiter's theory of diagnosis and belief revision are NP-complete problems.

> *Cancellation abduction Problems:* Cancellation abduction problems arise when constructing a composite hypothesis cancels the datum that another hypothesis may explain. Cancellation problem are defined formally as {D,H,m,p, e+ and e-}

where e+ is map from H to D indicating what data each hypothesis "produces" and e- is a map from H to D indicating what data each hypothesis "consumes." In other words, datum required by two hypotheses can cancel the expected effect on the datum from each other. Cancellation occurs when one hypothesis has a subtractive effect in the other.

The complexity measure for abduction when cancellation effects can be found in the domain is NP-complete for finding an explanation and as well as to find the best explanation.

A summary of the computational complexity of finding explanations is provided in Table 11. The condition of parsimony can be introduced to eliminate hypotheses choosing the best-small explanation with the constraint on more plausible hypotheses to less plausible ones. Adding this criterion in evaluating the best explanation for the independent abduction problems still leaves the problem complexity for finding the best explanation NP-hard. The task of finding the best-small explanation for ordered abduction problems becomes tractable when we assume that there exists an ordering of hypotheses by their plausibility. The problem of finding a best explanation for ordered hypotheses reduces to $O(n^2)$. However, finding an additional explanation under these conditions is still NP-hard.

Class of Problems	Condition to Achieve		
	Finding all Explanations	Finding an explanation	finding a best explanation
independent	NP	P	?
monotonic	NP	P	?
incompatibility	NP	NP	P
cancellation	NP	NP	P

Table 11: Computational complexity of finding explanations

The above highlights the point that only under the restrictions of no incompatibility relationships, cancellation relations and the individual hypothesis plausibility are computability for the abduction problem different from each other.

There have been efforts in the use of neural networks for abduction that follows the explanation models of abduction presented by [Bylander 1990]. [Ayeb 1998] claims that they have unified the algorithms to solve a class of abduction problems with neural network models. It is known that neural nets are NP-hard problems themselves.

3.3.3.1.3 Limitations of the Best Explanation Models of Abduction

[Thagard 1997] and [Magnani 2005] make the case that the best explanation model of abduction is only a partial model of abduction. They contend that equating explanation to deduction is not acceptable. The contention here is that these models are excluding the cases of abduction where hypotheses are formed and evaluated but do not provide deductive explanations. [Thagard 97] makes the case that the above models do not provide sufficient conditions for best explanations and often the parsimony conditions such as minimal subset for explanation are insufficient. He makes the case that additional causal relationships are required as in the case of probabilistic models of abduction such as belief nets [Pearl 1988][Peng 1990].

The other limitation is that best explanation models do not have hierarchy of hypotheses where the hypothesis serves as the effect for the hypothesis that represents the cause. While this approach can be incorporated in the Belief net model of abduction, it is not part of the deduction based models of abduction. The other important objection, in the case of Konolige's model, is that the hypothesis has to be consistent with the domain theory. This restriction of the model prevents the creation of new hypotheses as well incorporation of new hypotheses. Further, in areas where the extant domain theory is incomplete or the data leads to the creation of new domain theory, the restriction on creating explanations within the domain theory or only to a predetermined set of hypotheses will be limiting in creating the explanation for the new data in the old theory.

The restriction primarily placed by [Bylander 1990], in their deduction based model of abduction, insists on completeness of an explanation in consuming all the data and leaves out "partial explanations" as a possibility. This restriction prevents its use in most scientific and other realistic conditions. The other limitation is that all of these methods described until now are limited sentential representations.

3.3.3.2 Creative/selective Model

The creative/selective model is a kind of abduction that only generates plausible hypotheses. The selected hypothesis might be the best explanation. This process creates at least a partial explanation of the hypothesis. This form of abduction is more effective than the blind hypothesis generation. Selective abduction takes place more in the context of diagnostic tasks while creative abduction is hypothesised to take place in scientific discoveries [Magnani 2005]. In selective abduction, the task is to select a hypothesis from a set of pre-stored diagnostic entities while creative abduction adds new hypotheses to create a set of competing theories.

Abduction as seen in the creative/selective model is a holistic process where puzzling facts are fitted into coherent pattern of representations. The notion of coherence is characterized by maximization of constraint satisfaction [Thagard 1998]. They provide the following definition of coherence:

> Let E be a finite set of elements e_i and C be the set of constraints on E understood as a set (e_i, e_j) of pairs of elements E. C divides into C+, the positive constraints on E and C-, the negative constraints on E. With each constraint is associated a number w, which is the weight of the constraint. The problem is to partition E into two sets, A (accepted) and R (rejected), in a way that maximizes compliance with the following two coherence conditions:

1. if (ei, ej) is in C+, then ei is in A if and only if ej is in A;
2. if (ei,ej) is in C- then ei is in A is and only if ej is in R.

Let W be the weight of the partition, that is, the sum of weights of the satisfied constraints. The coherence problem is then to partition E into A and R in a way that maximizes W.

The elements are representation of cause and effects. This model of abduction allows for sentential and non-sentential (visual) forms of abduction. This model of coherence assumes that the relation between elements is an explanation and not a deduction. The decision concerning the explanatory hypothesis is one of maximizing the compliance of the two coherence conditions. The problem is computationally intractable, but there exist approximation algorithms that work quite well [Thagard 1997]. We have not evaluated these algorithms. In general, coherence abduction is based on maximizing the coherence by explaining as much as possible using positive constraints and being as consistent as possible using negative constraints.

3.3.3.3 Conclusions on Abduction

Abduction does not have a unique logic of its own. All formulations of abduction problems are NP-hard. Restrictions in the corpus of information in the domain are critical in deciding which form of abduction is the most suitable form to be used. A paper by [Bylander 1990] illustrates, through an example in medical diagnosis, that the formalism for abduction chosen is based on the nature of the problem. The general message in the development of metrics for abduction is that the goal of the problem being solved is critical in the choice of the abduction process. The problem of how the hypothesis layering is structured and the types of heuristics that are suited to the problem are required in the evaluation of the system. Abduction systems have the same dimensions of expressivity, computational complexity and completeness and soundness. Expressivity is different in this context, in contrast to deductive systems. What can be expressed is based on the type of statements that are allowable to decrease the measure of complexity from NP to P.

Further, the insistence on completeness and soundness are measures that are best suited to deductive systems and not abduction systems, except in the formulation of abduction as deduction by [Bylander 1990]. Completeness is a measure that is used when abduction is viewed as deduction or else it is disregarded. The other measure is the incorporation of new hypotheses or generation of new hypotheses both these require learning or revision of belief as in the case of belief nets as integral to the abduction process.

Evaluation of any abduction system would require several steps. First is the characterization of the model of abduction used, then identification of the upper bounds of complexity. Given these two measures of structure and complexity, restrictions or heuristics will be required to improve the measure. The measures themselves will need to be translated to metrics would require development of test cases that are average cases to provide a metric of algorithmic complexity in terms of the number of hypotheses, depth of the tree of the hypotheses and the number of elements (causal relations, data hypothesis relationships) used in the domain. The task of developing a metric is outside the scope of this document as the context of use is too general.

3.3.4 Analogical reasoning

Analogical reasoning is traditionally defined as similarity, resemblance, and correspondence, but it is also viewed as a comparison of dissimilar things with some similarity with respect to attributes and an inference can be based on that similarity [Sowa 2003].

3.3.4.1 Structure Mapping

One model of analogy assumes that representational structures already exist and the question of analogy is one of graph matching of two structures describing the object or phenomena to arrive at the conclusion of one thing being similar to the other [Gentner 1997] [Falkenheiner 1989]. A structure mapping engine (SME) has as inputs a base and target represented as two structured propositional representations. The propositional representations have the predicate/ argument structure. The predicate argument structure may include unary predicates that indicate features to represent attributes, relations that correspond to connections between predicates and higher order relations that correspond to connections between relations.

An SME for similarity-based retrieval [Gentner 1997][Forbus 2000], given a base and target, computes one or more mappings. The mappings are correspondences that align particular items in the base and the target domains. Further, they identify candidate inferences that are statements that hypothesized about the base that are expected to hold in the target implied by the correspondences of the items in the mapping.

The fundamental idea in structure mapping for analogy and similarity is that there exists a representational system that is explicit about the relational structure of the relationships. The representational structure is also explicit about the bindings between the objects and other relationships including causal dependencies across the domains of interest. The elements of the representation in this scheme are objects, object descriptors (attributes), functions (that express dimensional information), and relations between elements. In some cases, such as color of an object, a relation is represented as a function, relation or an attribute. The representation of a property will determine how it is processed.

Characterizing an analogy between two entities requires *structural consistency, relational focus,* and *systematicity.* Structural consistency is believed to exist when it observes parallel connectivity and one-to-one correspondence. Parallel connectivity is observed when matching arguments have matching predicates and one-to-one correspondence limits the matching of an element in one representation to exactly one entity in the other representation. Relation focus in an analogy focuses on common relationships and does require common object descriptions for it to work. The characteristic of systematicity in an analogy is achieved when an interconnected set of higher order relations are matched. The systematicity principle biases the preference for coherence and causal relations over a large number of unconnected relations.

In a structure mapping approach, once an alignment of structure is made, new inferences can be made from the analogy. Systematicity is an important principle as once the coherence and causal relations are mapped then new inferences in the target domain can be made because of the alignment of the base and the target domains. These new inferences in the target domain are candidate inferences that are verified in the target domain. Analogy and similarity are considered similar as the process of mapping between the target and base domains are similar. In the case of analogy, the correspondences of the inter-connected higher order relationships are critical without necessarily any correspondences between object descriptors. When there are detailed correspondences between the target and base domains then we are in the realm of literal similarities.

A structure mapping engine that implements the structure mapping process as a computational model uses a local-global alignment process to determine the structural alignment of the target and base domains. There are three steps in the computational model. The steps are: 1) local matches, where identical predicates and sub-predicates are matched in the two representations, 2) identifying structural consistencies. The first step produces typically inconsistent mappings with several many-to-one mapping. In this second step local matches are aggregated into one or few structurally consistent correspondences and 3) evaluation of structures, using the structurally consistent correspondences a structural evaluation of the correspondences are made using cascade-like algorithms in which evidence is passed down from predicates to their arguments. From the predicates connected to the common structure, candidate inferences are made in the target domain. An SME creates more than one interpretation for an analogy. A good explanation for an analogy can then be augmented using additional information from the context or other long term memory. This process is known as incremental mapping. [Falkenheiner 1989] show that the above SME operates in polynomial time, bounded by $O(n^2)$ where n is the number of nodes in the current knowledge base.

3.3.4.2 Copy-Cat Model of Analogical Reasoning and Its Extensions

The Copycat model, developed by Douglas Hofstadter and his students [Hofstadter 1993], claims that there no are explicit structures available for comparison ahead of time for the analogy to be derived but it requires the construction of the representation of the objects being compared from the senses, and argues for context dependent adaptable representations. In this model, the analogy problem is posed as, "If *abc* changes to *abd*, how does *iijjkk* change in an analogous way?" These problems do not have a unique answer and can have several defensible answers. The answer chosen by a human as best from the possible set of answers is often not the most obvious answer. The Copycat micro world, by generating a number of defensible answers, allows for experimentation in analogy making and high level perception.

To discover the analogy problem posed above, Copycat uses a non-deterministic stochastic processing distributed among large number of agents called *codelets*. These agents work on small parts of the analogy problem simultaneously without any high level co-ordination among the agents. The model's macroscopic behavior emerges from the actions of the microscopic events of the agents. The answer to the analogy problem *abc->abd; iijjkk->??*, is generated by the codelets working together to create a coherent mapping between the initial string *abc* and the target string *iijjkk* and also between the initial string *abc* and the modified string *abd*. In the creation of the mapping the codelets create hierarchical groups within the strings. The strings can be thought of as raw perceptual data that are grouped, say using the "sameness-groups" into *ii*, *jj* and *kk*. These are then arranged using what is called "successorship" group comprised of the above chunked groups. A mapping in Copycat consists of a set of Bridges between corresponding letters and groups depending on the role they play in the different strings. For example, a bridge between *a* in *abc* and *ii* in *iijjkk* is supported by, what are called concept mappings, *leftmost=>leftmost* and *letter => group*. These mappings represent the fact that the two elements in the strings compared are leftmost in their occurrence where one is letter and the other is a group. When the mappings are based on the identity of the element not preserved, i.e. the letter in the base string does not correspond to a letter in the target string but to a group as in the example, then they are called slippages. This allows Copycat to recognize superficially dissimilar situations.

Besides the mappings, for an analogy to be completed, Copycat requires *rules*. Rules are mappings between the initial string *abc* and the modified string *abd*. There are many ways for creating the mapping by identifying the type of change that is perceived from the initial string to the modified string. In the example above where *abc* is changed to *abd* one can write the rule to be "*change-rightmost letter to d*" or "*change the rightmost letter to its successor*." Use of these rules with the slippage "letter-group" on the target string will lead to *iijjdd* or *iijjll*. If we use the letter-letter bridge, we will end up with *iijjkl* or *iijjkd* as the result. To ensure that the answer *iijjll* is the answer, the rule to be applied will be *Change letter-category of rightmost group to successor*.

The Copycat model incorporates a measure of coherence called temperature. The temperature reflects the overall quality of the answer. This measure is crude and is one of the weaknesses of the Copycat model. The Copycat model does not preserve its internal underlying processes. This does not allow it to watch its own processes preventing it from providing explanation of its results. The other limitation is in the creation of rules as simple changes in string with a single letter changes are relatively easy but development of more complex rules was not included in Copycat.

In their subsequent work, named Metacat, Hofstadter and students extend Copycat to include the ability to create limited explanation of answers in terms of the strengths and weaknesses. Details of this algorithm can be found in [Marshall 1996]. There is no study of complexity of these processes in the literature.

3.3.4.3 Hybrid Model of Analogy

Another view argues for inclusion of both forms of analogical reasoning. [Sowa 2003] proposes a model (Vivomind) that combines both forms, taking the position that for a comprehensive theory of cognition, analogical reasoning must include both the bottom-up, perception-based approaches advocated by Hofstadter and the SME approach advocated by Forbus and his co-workers. In the Vivomind approach, analogies are found using SME. It also has algorithms that are used to build the structures used in the mapping for analogies. It creates structures using Sowa's conceptual graphs by parsing natural language. The method is extended to include sensory data from percept-like patterns in scene recognition. It is claimed that the Vivomind Analogy Engine (VAE) can find analogies in time $O(nlogn)$. VAE uses conceptual graphs created from the analysis of any type of source: natural language, programming language, and any information represented as graphs.

The VAE method consists of three steps:

1. match type labels, which compares nodes that have identical labels, labels that related by type-subtype hierarchy.
2. match subgraphs with possibly a different label. The match stops when two graphs that are isomorphic are produced and when they can be made isomorphic by combining nodes in the graphs.
3. if the steps 1 and 2 fail, then a search for transformations that relate the subgraphs of one graph to the subgraphs of the other can is identified.

The use of these steps can be independent or in combination.

4 User Interfaces

Although Application Program Interfaces (APIs) are becoming increasingly data-centric, for instance, increasingly more commonly defined via the eXtensible Markup Language (XML) [W3C 2004a], user-interfaces suffer a schism. Specifically, implementations indeed use data-driven specifications; however, testing of such interfaces is still a largely ad hoc art. Metrics are similarly split. Although back-end metrics exist to indicate whether interfaces conform to basic standards, see Section 4.2, front-end metrics are far from any clear and useable level of standardization, see Section 4.4. To a great extent, this is reflected by highly specialized user interface testing software that is difficult to extend when faced with new conditions. It is not surprising that human users still remain responsible for the detection of the majority of errors in user interfaces throughout the industry.

4.1 User Interface Types

We categorize user interfaces into several different types. These types lend themselves to widely differing types of testing methods and metrics.

    Character-string
    Character-graphic
    Bit-mapped
    Physical

Character-string interfaces arise in traditional printer-style applications. These are generally tested using regular expressions and tools such as Expect [Libes 1995]. Character-graphic applications rely on "dumb terminal" applications are typically tested by screen snapshots or using a terminal proxy that can be queried by row and line, again, augmented with regular expressions.

Bit-mapped devices use a variety of techniques including optical character recognition (OCR), frequently driven by hints using both format and screen location [Wikipedia 2005]. Note that this technology can be applied subversively and for that reason is becoming increasingly useless in security applications that expect subversive use. Examples of this can be found at websites that present the user with intentionally distorted bit-mapped pass phrases [Yahoo 2005].

4.2 Front-end Methods and Metrics

Front-end methods refer to methods that measure activity at the front of the interaction – literally, the front of the screen. For example, comparing screen shots to known standards is a simple basis for a user interface metric.

Actual metrics are much more sophisticated, spanning a large range of techniques and specifications:
    Regular-expression string matching
    Graphic matching
    Graphic parsing and matching

4.3 Approaches

The approaches to front-end methods include:
    Simulated screen capture
    True screen capture
    Physical screen capture

Front-end methods are particularly adept at thorough user-experience testing. Software can detect the condition in which user interfaces fail from the human point of view. The drawback of front-end approaches is they provide little information about the implementation and can thus make software discrepancies difficult to locate and fix despite the knowledge of their existence.

4.4 Back-end Methods and Metrics

Back-end methods refer to methods that use artificial metrics "behind the scenes" such as observing the stream of data rather than its visual result. [Hampson 2005]
    Proxy events
    Parsed streams

Such methods have advantages and disadvantages compared to front-end testing. The primary advantages of back-end testing are that user interface objects can generally be detected at a higher-level. For example, a single protocol element may indicate the existence of a box or even better, a warning popup. In contrast, detecting this from the front-end would require detecting the box, that it is a warning (perhaps via an icon), and detecting the text and its meaning.

Back-end testing has disadvantages as well. It can sometimes be more difficult to interpret protocols than do front-end analysis. For example, an icon that is bit-mapped onto the screen provides little information at the protocol level but may render itself to trivial analysis via OCR.

A more fundamental difficulty of back-end analysis is that it requires hooks into the software or network that are sometimes unavailable or unreliable.

4.5 Physical Methods

Physical methods rely on human observance and interaction. In this case, an observer may interact with the software or may simply observe another user interacting. A log is written by the observer and used to measure the software. Obviously, this is a last resort but oftentimes the only practical solution.

The physical approach enables more likely knowledge of intent than any alternative. For example, if a user unsuccessfully searches for a particular menu item, automated approaches cannot detect the inability of the search. It may be implied by numerous traversals but only by witnessing the user can it be clear.

Some physical tests also include user interviews (either automated or not) in which users are queried for their thoughts. This can also be helpful in user interface analysis.

4.6 Products and Toolkits

There is a wide array of toolkits and products for the development of metrics of user interfaces. Some examples include:

- Front-end and back-end testing are illustrated by Morae [TechSmith 2005] and Glassbox [Cowley 2005]. Both of these systems record a chronical of events including those occurring both back-end and front-end methods. This includes low-level events such as mouse clicks and screen snapshots to audio and video recordings of users.

- Behind-the-scenes testing are illustrated by NetworkTester [Agilent 2005] and TestPerspective Load Test [Keynote 2005]. Both of these systems focus on back-end methods with emphasis on correctness and loading.

Lengthy lists of such user-interface testing software and technology can be found at a variety of websites such as softwareqatest.com [Hower 2005]. However, no sites exist that publish broad comparative evaluations of these products.

4.7 Representations and Standards

There are no universal representations and standards for the purpose of user interface metrics. What exists are fragmented by fields and even within fields include numerous representations. Approaches fall into two types:
- Specification-based
- Programmatic

Specification-based approaches to date are highly software-specific. These include such elements as screen snapshots and fuzz. Screen snapshots may be literally rendered (as a bitmap) or as a composition of elements. Fuzz refers to the level of difference that may be present between the specification and the actual user interface. For example, a screen snapshot may differ only by a changing date with each test. Clearly, accommodation has to be made for this and other changes.

Programmatic approaches generally use a script-based specification such as Tcl, Python, or Java [Welch 2003][Lutz 2003][Flanagan 2005]. In these systems, a script describes the interaction using if-then-else statements, procedures, and anything else, algorithmically.

4.8 Comparisons

There is a wide array of toolkits and products for the development of metrics of user interfaces. It is important to recognize that they cover a multi-dimensional spectrum of power, flexibility, ease-of-use as well as other attributes.

For relatively simple interfaces such as character streams and character-graphic interfaces, solutions are well characterized by programmatic specifications [Libes 1995] [Savoy 2005].

Generally, tools that are more powerful are correspondingly more difficult to use. Metrics that are more specific are more brittle. This is particularly problematic when dealing with interfaces intended for humans who readily use common sense to interpret information that easily befuddle some of the best common-sense reasoning systems.

5 Software Interfaces

In the real world, reasoning systems are not deployed as monolithic software applications, but instead as parts of software systems where the reasoning function is one of multiple components. For instance, a decision support system may use a rule-based inference engine to guide recommendations. Because reasoning tools need to interoperate with other software, it is essential that they provide interfaces for doing so. But how can we assess the quality of a reasoning tool's software interface?

We begin by enumerating the various ways in which reasoning systems can interact with other software components and assessing which methods of interaction are most important. Then we discuss the value of integration standards for ensuring reasoning tool interoperability. In the course of our discussion, we provide some real world examples of software interfaces for reasoning systems.

5.1 Methods of Interaction

Methods of interaction between software applications are becoming increasingly data-centric. Although application program interfaces are still important, data exchange formats are being emphasized more and more. This is because of the widespread growth of distributed software systems relying on the Internet for transporting bits between nodes, and on XML for providing data structures that can be interpreted by the nodes involved in an exchange. Such systems usually use the Representational State Transfer (REST) architectural style [Fielding 2000] [Costello 2005]. In a "RESTful" design, conceptual entities are identified as *resources*. Each resource is retrievable using a Uniform Resource Locator (URL). The resources may be represented as HyperText Markup Language (HTML) or XML documents, or they may be images, plain text, binary files, etc. If the resource is HTML or XML, it may contain URLs linking to other resources. By clicking on these links, a user causes the system to change state.

A key characteristic of REST is that it is highly data-centric. The communication protocols of the Web are simple. REST's power and flexibility comes from representations of resources. Because XML allows application developers to define their own tag set vocabularies, an XML resource can be arbitrarily detailed. XML representations have other advantages as well. Many free and low-cost software tools can process XML, and most Web browsers have built-in XML processors. Another advantage of XML is its programming language independence. Because XML is programming language-neutral, and because XML processing tools are available in most commonly used programming languages, a single XML vocabulary can serve the needs of just about any software environment.

XML's ubiquity, extensibility, and programming language independence make it an attractive representation method, even for software applications not employing the REST paradigm. Increasingly, exchange formats, configuration languages, and the like are represented as XML. XML has in effect become a universal syntax for data exchange.

5.1.1 Data Exchange

Given the growing popularity of data-centric software development, and the common use of XML as a language for specifying data formats, it follows that it is useful for a reasoning tool to be able to import from and export to an XML data exchange format. The format should be well documented so that software integrators can easily write code to create and/or process data in the format. A key factor in evaluating a reasoning system's software interface is its ability to import from and export to standard formats (ideally XML formats) for knowledge bases, queries, or query results.

Fortunately there are several standard XML formats for representing reasoning tool data. One example is the Description Logic Interface (DIG) [Bechhofer 2003]. DIG is an XML language for expressing "ask" and "tell" description logic (DL) expressions. DL reasoners supporting DIG may be used interchangeably in a software application calling for DL inferencing. For example, the Protégé OWL plug-in works with any DIG-compliant reasoner.

XML languages for representing knowledge bases include OWL [W3C 2004b], a standard for representing ontologies, and RuleML [RuleML 2005], an XML format for representing production rules.

Although an XML exchange format is highly desirable, sometimes it makes sense to use a non-XML exchange mechanism. For instance, highly expressive knowledge representation languages such as first order logic and Prolog have non-XML syntax that is well understood and is less verbose than an XML syntax would be. Software tools are available to process these non-XML syntaxes. Therefore, there is not much point in creating a new XML syntax.

5.1.2 Application Program Interfaces (APIs)

Although the trend in software integration is toward XML and data formats, there is still a role for application program interfaces. An API is a set of classes or function calls exposing a subset of a software system's capabilities. The subset should ideally be small enough to be easy to understand, but large enough to support most tasks that a programmer would want to implement. Unlike exchange formats, APIs are usually specific to a particular programming language. Therefore, an API can be specifically tailored to the features of the programmer's desired language. However, if the developer wants to program in a different language, another API is needed.

Most reasoning systems have APIs. A recent trend is the notion of a modular "plug-in" architecture. These development environments make it easy to add new capabilities to an existing software system. A good example is the Protégé ontology editor [Protégé]. Thanks to the extensibility afforded by its plug-in API, numerous Protégé add-ons have been implemented.

Yet another potentially useful approach to integrating a reasoner with a user interface is an API allowing user interface programming to be done using the reasoner. SWI Prolog [Wielemaker 2005] uses this approach, providing a mechanism for specifying user interface operations as Prolog [Wielemaker 2002].

5.2 The Value of Standards

What do we mean by a "standard" in the context of reasoning system integration? A standard may be one of at least three types [Peak 2004]:

Open Standards provide an agreement that people make so that products and systems made by different entities can function together. Open standards are not software applications; rather they specify how data should be represented. Open standards are developed by consensus among parties with a shared interest in integration. Groups developing open standards range from official organizations like the International Organization for Standardization, or ISO (http://www.iso.ch), to small vertical industry groups.

The Process Specification Language is an example of an open standard [ISO 2005]. It is developed by ISO, with the help of industry and academia. XML is also an open standard, even though it is being developed outside ISO. XML is developed by the World Wide Web Consortium (W3C, http://www.w3.org). It is common for an open standard to originate from a successful research project. For example, the Prolog language, which originated from university research, subsequently became an ISO standard.

Industry Standards are popular technologies that are not open or democratically managed by a group of users. The Java™ technology is an example. Many companies are involved in the Java Community Process (http://jcp.org), but because one company exerts a tremendous amount of control over the process, Java is an industry standard, not an open standard.

De facto Standards are widely used because of their value or association with other technologies, and not necessarily because they were produced by a standards organization. A commercial software product may be a *de facto* standard because of its wide adoption. The Microsoft Windows operating system is a *de facto* standard for personal computers. *De facto* standard status does not mean that there are no alternatives, but that such alternatives are rarely used.

Besides the three types of standards, there is open source software. Open source software is not necessarily an open standard. Open source refers to software source code that is available to the general public and does not have licensing restrictions that limit use, modification, or redistribution under the same terms as the license of the original software[75]. The GNU/Linux operating system (http://www.linux.org) and Eclipse software development environment (http://www.eclipse.org) are examples of open source technologies.

All other characteristics being equal, a more open software interface technology is preferable to a less open one. Best of all is an open standard, where the interface cannot change without consensus from a community of users. Next best is an industry standard interface with a published specification. Although the interface can change without consent from the user community, at least the users have access to documentation.

---

[75] See http://www.opensource.org/docs/definition_plain.php for a more detailed definition of "open source".

5.3 Software Interface Quality

So far we have emphasized the desirability of standardization and the desirability of favoring widely used description and implementation technologies. However, these attributes alone do not fully determine the quality of a software interface. It is important that the representation of the interface be consistent and complete, yet also easy to understand by developers.

Consistency, completeness, and understandability are characteristics of the software interface's underlying *information model*. This information model is akin to any other information model, except that it is tailored for exchange rather than direct data manipulation. A software interface's information model may be expressed in one or more of a number of languages, such as XML or UML. Determining the quality of an information model amounts to measuring its syntactic and its semantic properties. Syntactic properties include characteristics such as naming conventions for information objects as well as presentation characteristics dependent on the modeling language used. These properties can be measured in terms of conformance to naming and design standards, such as XML naming and design rules (see http://xml.coverpages.org/ndr.html).

Semantic quality of an information model is more difficult to measure. Traditionally, semantic quality is determined by domain experts who review the model and assess how closely it aligns with the expert's view of the data being exchanged. This process is labor-intensive and can be subjective.

6 Conclusion and Future Work

A review of the literature on evaluating reasoning systems reveals that it is a very broad area with wide variation in depth and breadth of research on metrics and tests. Consolidation is hampered by nonstandard terminology, differing methodologies, scattered application domains, unpublished algorithmic details, and the effects of domain content and context on the choice of metrics and tests. The field of information metrology, which applies to reasoning as a kind of information processing, is still emerging from ad hoc experience in evaluating narrow kinds of information systems.

This report begins to bring order to the area by categorizing reasoning systems according to their capabilities. These categories give metrics for determining if a particular system is performing the required kinds of reasoning. The capabilities are analyzed along several dimensions, including representation languages, inference, and user and software interfaces. The report introduces information metrology, model theory, and inference to facilitate understanding of the reasoning categories presented.

Applying the results of this report to evaluation of specific reasoning systems requires the following steps:

1. Further refinement of reasoning categories as needed to support development of test cases of interest to reasoning system users.
2. Development of generic test cases for reasoning categories. Generic tests are independent of the application, reasoning tool, and computational platform.
3. Development of specific test cases that are dependent on application, reasoning system tool, and computational platform, based on the generic test cases above.
4. Application of steps 1-3 above (refine categories, develop generic tests, develop specific tests) to user interface and software interface capabilities.

These steps can only be completed in partnership with users of reasoning systems and reasoning tool providers.

7 Acknowledgements and Disclaimer

The U.S. National Security Administration funded this work under Contract 712065 for the High Performance Evidentiary Reasoning Program.

Thanks to Jeffery Smith ,William (Bill) Andersen, and Peter Denno for their reviews.

Commercial equipment and materials might be identified to adequately specify certain procedures. In no case does such identification imply recommendation or endorsement by the U.S. National Institute of Standards and Technology, nor does it imply that the materials or equipment identified are necessarily the best available for the purpose.

8 References

[Agilent 2005] Aligent, "NetworkTester," http://advanced.comms.agilent.com/networktester, 2005.

[Alferes 2003] Alferes, J., Damasio, C., Pereira, L., "Semantic Web Logic Programming Tools," Workshop on Principles and Practice of Semantic Web Reasoning, at 19th Internatioinal Conference on Logic Programming (ICLP03) 2003.

[Allen 1983] Allen, J.F., "Maintaining knowledge about temporal intervals," Communications of the Association for Computing Machinery, 26:11, pp. 832-843, 1983.

[Alonso 1994] Alonso, F., Mate, L., Juristo, N., Munoz, P., Pazos, J., "Applying Metrics to Machine-Learning Tools: A Knowledge Engineering Approach," AI Magazine 15:3, pp. 63-75, http://citeseer.ist.psu.edu/220394.html, Fall 1994.

[Alvaro 1990] Alvaro, R., "Induction as a Solution for Knowledge Acquisition: AQ Algorithm," Master's thesis, Facultad de Informática, Universidad Politécnica de Madrid, 1990.

[An 2001] An, A., Cercone, N., "Rule quality Measures for Rule Induction Systems: description and evaluation," Computational Intelligence, 17:3, pp. 409-424.

[Antoniou 1997] Antoniou, G., Nonmonotonic Reasoning, MIT Press, 1997.

[Apt 1987] Apt, K., Blair, H., Walker, A., "Towards a theory of declarative knowledge," in Foundations of Deductive Databases and Logic Programming, Minker, J. (ed.), pp. 89-142, Morgan Kaufmann, 1987.

[Argonne 2004] Otter: "Otter: An Automated Deduction System," Argonne National Laboratory, http://www-unix.mcs.anl.gov/AR/otter, 2004.

[Arvind 1987] Arvind, V., Biswas, S. "An O(n2) algorithm for the satisfiability problem of a subset of propositional sentences in CNF that includes all Horn sentences," Information Processing Letters 24, pp. 67-69, 1987.

[Ayeb, 1998] Ayeb, B, Wang, S and Ge, J., "A unified model of Abductive reasoning," IEEE transactions on Systems Man and Cybernetics, 1998.

[Bachmair 1995] Bachmair, L., Ganzingery, H., Lynchz, C., Snyderx, W., "Basic paramodulation," Information and Computation Journal, 121:2, pp. 172-192, September 1995.

[Baral 1994] Baral, C., Gelfond, M., "Logic Programming and Knowledge Representation," Journal of Logic Programming, 19:20, pp. 73-148, 1994.

[Baumgartner 2004] Baumgartner, P., Mediratta, A., "Improving Stable Models Based Planning by Bidirectional Search," International Conference on Knowledge Based Computer Systems (KBCS), Hyderabad, India, December 2004.

[Bechhofer 2003] Bechhofer, S., "The DIG Description Logic Interface: DIG 1.1. University of Manchester," http://dl-web.man.ac.uk/dig/2003/02/interface.pdf, February 2003.

[Bock 2000] Bock, C., "Goal-driven Modeling," Journal Of Object-Oriented Programming, 13:5, pp. 48-50, September 2000.

[Bock 2001] Bock, C., "Goal-driven Modeling, Part II," Journal Of Object-Oriented Programming, 13:11, pp. 25-28, March 2001.

[Boolos 1980] Boolos, G. Jeffrey, R., Computability and Logic, 2nd ed., Cambridge University Press, 1980.

[Boolos 1993] Boolos, G., The Logic of Provability, Cambridge University Press, 1993.

[Brieman 1984] Breiman, L, Friedman, J.H., Olshen, R.A., Stone, C.J., Classification and Regression Trees, Belmont, 1984.

[Broxvall 2003] Broxvall, M., Jonsson, P., "Point algebras for temporal reasoning: algorithms and complexity," Artificial Intelligence, 149:2, pp.179-220, October 2003.

[Bylander 1990] Bylander, T, Allemang, D., Tanner, M.C., Josephson, J, "The computational complexity of Abduction," Artificial Intelligence Journal, 1990.

[Cadoli 1993] Cadoli, M., Schaerf, M., "A Survey of Complexity Results for Nonmonotonic Logics," Journal of Logic Programming, 17:2-4, pp.127-160, 1993

[Calvenese 2001] Calvenese, D., De Giacomo, G., Nardi, D., Lenzerini, M., "Reasoning in expressive description logics," in Handbook of Automated Reasoning, A. Robinson and Voronkov, A. (eds.) Elsevier Science Publishers. 2001.

[Carnahan 1997] Carnahan L., Carver, G., Gray, M., Hogan M., Hopp T., Horlick, J., Lyon, G., Messina, E., "Metrology for Information Technology," StandardView, 5:3, pp. 103-109, September, 1997.

[Carnap 1947] Carnap, R., Meaning and Necessity, University of Chicago Press, 1947.

[Carson 1967] Carson, D., Robinson, G., Shalla, Wos, L. "The Concept of Demodulation in Theorem Proving," Journal of the Association for Computing Machinery, 14:4, pp. 698-709, October 1967.

[Castro 2005] Castro, L., Swift, T., Warren, D., "XASP, Answer Set Programming with XSB and Smodels," http://xsb.sourceforge.net/packages/xasp.pdf, 2002

[Cestnik, 1987] Cestnik, B., "ASSISTANT PROFESSIONAL, A Software Tool for Inductive Learning of Decision Rules. System User Manual," Edvard Kardelj University, Ljubljana, Yugoslavia, 1987.

[Chandra 1985] Chandra, A., Harel, D., "Horn clause queries and generalizations," Journal of Logic Programming, 25:1, pp. 1-15, 1985.

[Chang 1973] Chang, C. Keisler, H., Model Theory, North Holland, 1973.

[Chen 1993] Chen, W., Kifer, M., Warren, D., "HILOG: A Foundation for Higher-Order Logic Programming," Journal of Logic Programming 15, pp. 187-230, 1993.

[Chen 1996] Chen, W., "Extending Prolog with nonmonotonic reasoning," Journal of Logic Programming 27:2, pp. 169-183, 1996.

[Clark 1978] Clark, K., "Negation as Failure," in Logic and Databases, Gallaire, H., Minker, J. (eds.), pp293-322, Plenum Press, 1978.

[Clark 1989] Clark, P., Niblett, T., "The CN2 algorithm," Machine Learning, 3:4, pp. 261-284, 1989.

[Clocksin 2003] Clocksin, W., Mellish, C., Programming in Prolog : Using the ISO Standard, Springer, 2003.

[Cohen 1982] Cohen, P., Feigenbaum, E., The Handbook of Artificial Intelligence, vol. 3, pp. 383-451, William Kaufamann,1982.

[Colmerauer 1993] Colmerauer, A., Roussel, P., "The birth of Prolog," Association for Computing Machinery SIGPLAN Notices, v.28 n.3, p.37-52, March 1993, reprinted with presentation and transcript in "History of programming languages-II," Association for Computing Machinery Press, New York, NY, 1996.

[Costello 2005] Costello, R., Building Web Services the REST Way, xFront, http://xfront.com/REST-Web-Services.html, 2005.

[Cowley 2005] Cowley, P., Nowell, L., Scholtz, J., "Glass Box: An Instrumented Infrastructure for Supporting Human Interaction with Information," Proceedings of the Proceedings of the 38th Annual Hawaii International Conference on System Sciences (HICSS '05), 9, 2005.

[Cresswell 1991] Cresswell, M., "In Defence of the Barcan Formula," Logique et Analyse, 135-136, 271-282, 1991.

[Dieterich 1981] Dieterich, T., and Michalski, R., "Inductive Learning of Structural Descriptions: Evaluation criteria and comparative review of selected methods," Artificial Intelligence, 16, pp. 257-294, 1981.

[Ding-Zhu 2000] Ding-Zhu D., Ko, K., Theory of Computational Complexity, Wiley-Interscience, 2000.

[Doets 1994] Doets, K., From Logic to Logic Programming, MIT Press, 1994.

[Donini 1996] Donini, F., Lenzerini, M., Nardi, D., Schaerf, A., "Reasoning in description logics," in Principles of Knowledge Representation, Brewka, G. (ed.), pp. 193-238, CSLI Publications, 1996.

[Donini 1997] Donini, F., Lenzerini, M., Nardi, D., Nutt, W., "The complexity of concept languages," Information and Computation, 134, pp. 1-58, 1997.

[Dowling 1984] Dowling, H., "Linear-time algorithms for testing the satisfiability of propositional Horn formulae," Journal of Logic Programming 3, pp. 267-284, 1984.

[East 2001] East, D., Truszczynski, M., "Propositional satisfiability in answer-set programming," Lecture Notes in Computer Science," 2174, pp. 138-153, Springer Verlag, 2001.

[Ebbinghaus 1999] Ebbinghaus, H., Flum, J., Finite Model Theory, Springer-Verlag, 1999.

[Eiter 2003] Eiter, T., Faber, W., Leone, N., Pfeifer, G., Polleres, A., "A logic programming approach to knowledge-state planning, II: The DLV System," Artificial Intelligence, 144:1-2, pp. 157-211, March 2003.

[Enderton 1972] Enderton, H., A Mathematical Introduction to Logic, Academic Press, 1972.

[Fagin 2003] Fagin, R., Halpern, J., Moses, Y., Vardi, M., Reasoning about Knowledge, MIT Press, 2003.

[Falkenheiner 1989] Falkenhainer, B, Forbus, K., Gentner, D., "The Structure Mapping Engine: Algorithms and Examples, Artificial Intelligence," 41, pp.1-63, 1989.

[Feldman 1991] Feldman, Y., Rich, C., "Pattern-directed invocation with changing equations," Journal of Automated Reasoning, 7:3, pp. 403-433, September 1991.

[Fernández 1990] Fernández, A., "Comparative Study of Prune Methods for Decision Tree Induction Algorithms," Master's thesis, ETSI Telecomunicaciones, Universidad Politécnica de Madrid, 1990.

[Fielding 2000] Fielding, R.T., "Architectural Styles and the Design of Network-based Software Architectures," Ph.D. Thesis, University of California, Irvine, http://www.ics.uci.edu/~fielding/pubs/dissertation/top.htm, 2000.

[Filman 1990] Filman, R., Bock, C., Feldman, R., Singer, J., Treitel, R., "New Generation Knowledge-System Tools," Defense Advanced Research Projects Agency, RADC-TR-89-383, April 1990.

[Fisher 1987] Fisher, D.H., "Knowledge acquisition via incremental conceptual clustering," Machine Learning, 2, pp.139-172, 1987.

[Fitting 1998] Fitting, M., Mendelsohn, R., First Order Modal Logic, Kluwer, 1998.

[Fitting 2002] Fitting, M., "Fixpoint Semantics for Logic Programming A Survey," Theoretical Computer Science, 278:1-2, pp. 25-51, May 2002.

[Flach 2001] Flach, P., "On the state of Art of Machine Learning: a personal review," Artificial Intelligence, 13:1/2, pp. 199-222, September 2001.

[Flach, 1998] Flach, P., "The Logic of Learning: A Brief Introduction to Inductive Logic Programming," Proceedings of the CompulogNet Area Meeting on Computational Logic and Machine Learning, pp. 1-17, University of Manchester, http://www.cs.brid.ac.uk/~flach, June 1998.

[Flanagan 2005] Flanagan, D., Java In A Nutshell, 5th Edition, O'Reilly Media, March 2005.

[Flum 1985] Flum, J., "Characterizing logics," in Model-Theoretic Logics, Barwise and Feferman (eds.), Springer-Verlag, 1985.

[Forbus 2000] Forbus, K., "Exploring analogy in the large," in Gentner, D., Holyoak, K., Kokinov, B. (eds.) Analogy: Perspectives from Cognitive Science, MIT Press, 2000.

[Gabby 1998] Gabbay, D., Hogger, C., Robinson, J., Handbook of Logic in Artificial Intelligence: Logic Programming, Oxford University Press, 1998.

[Gelfond 1991] Gelfond, M., and Lifschitz, V., "Classical Negation in Logic Programs and Disjunctive Databases," New Generation Computing, 9:3-4, pp.365-386, 1991.

[Gelfond 2002] Gelfond, M., Leone, N., "Logic Programming and Knowledge Representation - the A-Prolog Perspective," Artificial Intelligence, 138:1-2, pp. 3-38, 2002.

[Genesereth 1987] Genesereth, M., Nilsson, N., Logical Foundations of Artificial Intelligence, Morgan Kaufman, 1987.

[Gentner 1997] Gentner, D, Markman, A, "Structure Mapping in Analogy and Similarity," American Psychologist, 1997.

[Giunchiglia 2005] Giunchiglia, E., Maratea, M., "On the relation between Answer Set and SAT procedures (or, between CMODELS and SMODELS)," Proceedings of the 21st International Conference on Logic Programming (ICLP 2005) October 2005, to appear Lecture Notes in Computer Science.

[Golumbic 1993] Golumbic, M., Shamir, R., "Complexity and algorithms for reasoning about time: a graph-theoretic approach," Journal of the Association for Computing Machinery 40:5, pp. 1108-1133. DOI= http://doi.acm.org/10.1145/174147.169675, November 1993.

[Gonzalez 1998] Gonzalez, A.J., Xu, L., Gupta, U.M., "Validation Techniques for Case-Based Reasoning Systems," IEEE transactions on Systems Man and Cybernetics - Part A, Systems and Humans, 28:4, pp.:465-477, 1998.

[Gray 1999] Gray, M., "Applicability of Metrology to Information Technology," Journal of Research of the National Institute of Standards and Technology, 104:6, pp. 567-578, November/December 1999.

[Griener 2001] Greiner, R., Darken, C., Iwan Santoso, N., "Efficient Reasoning," Association of Computing Machinery, Computing Surveys, 33:1, pp. 1-30., http://www.cs.ualberta.ca/~greiner/PAPERS/EffReason.ps, March 2001.

[Grosof 1999], B., Labrou, Y., Chan, H., "A Declarative Approach to Business Rules in Contracts: Courteous Logic Programs in XML," Proceedings of the 1st Association for Computing Machinery Conference on Electronic Commerce (EC99), Wellman, M (ed.), Association for Computing Machinery Press, November 1999.

[Grosof 2003] Grosof, B., Horrocks, Ian., Volz, R., Decker, S., "Description Logic Programs: Combining Logic Programs with Description Logic," Proceedings of the Twelfth International World Wide Web Conference (WWW-2003), pp. 48-57, Association for Computing Machinery, 2003.

[Guarino 2002] Guarino, N., Welty, C., "Evaluating Ontological Decisions with OntoClean," Communications of the Association for Computing Machinery, 45:2, pp. 61-65, Association for Computing Machinery Press, 2002.

[Gyftodimos 2004] Gyftodimos, E., Flach, P., "Hierarchical Bayesian Networks: An Approach to Classification and Learning for Structured Data," Proceedings of Methods and Applications of Artificial Intelligence, Third Hellenic Conference on Artificial Intelligence (SETN 2004), Vouros, G., Panayiotopoulos, T., (eds.), pp. 291-300, Springer, http://www.cs.bris.ac.uk/Publications/pub_info.jsp?id=2000217, May 2004.

[Haarslev 2003] Haarslev, V., Möller, R., "Racer: A Core Inference Engine for the Semantic Web," Proceedings of the 2nd International Workshop on Evaluation of Ontology-based Tools (EON2003), at 2nd International Semantic Web Conference ISWC 2003, pp. 27-36, October 2003.

[Hall 2001] Hall, M., Zeleznikow, J, "Acknowledging the Insufficiency in the Evaluation of Legal Knowledge Based systems: Strategies Towards Broad-based Evaluation Tool," International Conference on Artificial Intelligence and Law 2001, pp. 147-156, 2001.

[Hampson 2005] Hampson, E., and Cowley, P., "Instrumenting the Intelligence Analysis Process," Proceedings of the 2005 International Conference on Intelligence Analysis, https://analysis.mitre.org/proceedings/Final_Papers_Files/141_Camera_Ready_Paper.pdf, 2005.

[Hewitt 1969] Hewitt, C., "PLANNER: A Language for Proving Theorems in Robots," First International Joint Conference on Artificial Intelligence, pp. 295-301, 1969.

[Heymans 2004] Heymans, S., Van Nieuwenborgh, D., Vermeir, D., "Semantic web reasoning with conceptual logic programs," Proceedings of the Third International Workshop on Rules and Rule Markup Languages for the Semantic Web (RuleML 2004), pp. 113-127, 2004.

[Hietalahti 2000] Hietalahti, M., Massacci, F., Niemela, I., "DES: a Challenge Problem for Nonmonotonic Reasoning Systems," Proceedings of the 8th International Workshop on Non-Monotonic Reasoning, April 2000.

[Hodges 1993] Hodges, W., Model Theory, Cambridge University Press, 1993.

[Hofstadter 1994] Hofstadter, D., and Mitchell, M., "The Copycat project: A model of mental fluidity and analogy-making," In Holyoak, K., Barnden, J. (eds.), Advances in Connectionist and Neural Computation Theory, Volume 2: Analogical Connections, pp. 31-112, Ablex, 1994.

[Horrocks 1999] Horrocks I., Sattler, U., Tobies, S., "Practical reasoning for expressive description logics," Proceedings of 6th International Conference on Logic for Programming and Automated Reasoning (LPAR-99), reprinted in Lecture Notes in Artificial Intelligence, 1705, 1999.

[Hower 2005] Hower, R., "Software QA/Test Resource Center," http://www.softwareqatest.com/qatweb1.html, 2005.

[Hughes 1984] Hughes, G., Cresswell, M., A Companion to Modal Logic, Methuen, 1984.

[Hughes 1996] Hughes, G., Cresswell, M., A New Introduction to Modal Logic, Routledge, 1996.

[ISO 1995] ISO/IEC 13211-1:1995, Information technology - Programming languages - Prolog - Part 1: General core, 1995.

[ISO 2005b] ISO 18629:2005, Process Specification Language, http://www.tc184-sc4.org/SC4_Open/SC4_Work_Products_Documents/PSL_(18629), 2005.

[ISO 2005a] ISO/IEC JTC 1/SC 32 (2005) ISO/WD 24707 -- Common Logic -- Framework for a family of logic-based languages, 2005

[Ignizio 1990] Ignizio, J., An Introduction to Expert Systems: The Development and Implementation of Rule Based Expert Systems, Mcgraw-Hill, 1990.

[Jammer 1997] Jammer, M., Concepts of Mass in Classical and Modern Physics, Dover, 1997

[Kamber, 1997] Kamber, M., Winstone, L., Gong, W., Cheng, S., and Han, J., "Generalization and decision tree induction: efficient classification in data mining," Proceedings of the 7th international Workshop on Research Issues in Data Engineering (RIDE '97) High Performance Database Management For Large-Scale Applications, IEEE Computer Society, Washington, DC, 111, April 1997.

[Keisler 1971] Keisler, H.J., Model Theory for Infinitary Logic: Logic with Countable Conjunctions and Finite Quantifiers, North-Holland, 1971.

[Keynote 2005] Keynote, Inc., http://www.keynote.com, 2005

[Konolige 1992] Konolige, K., "Abduction versus closure in Causal Theories," Artificial Intelligence, 53:2-3, pp. 255-272, 1992.

[Kowalski 1995] Kowalski, R., "Logic without Model Theory," in What is a Logical System?, Gabbay, D., (ed.), pp. 35-71, Oxford University Press, 1994.

[Kowalski 2002] Kowalski, R., "Directions for Logic Programming," Computational Logic: Logic Programming and Beyond, Essays in Honour of Robert A. Kowalski, Part I, Kakas, C., Sadri, F. (eds.), Lecture Notes In Computer Science, 2407, pp 26-32, Springer, 2002.

[Krokhin, 2003] Krokhin, A., Jeavons, P., Jonsson, P., "Reasoning about temporal relations: the tractable subalgebras of Allen's interval algebra," Journal of the Association for Computing Machinery, 50:5, pp.591-640, September 2003

[Kumar 2005] Kumar, S., "Contributions to Algorithmic Techniques in Automated Reasoning about Physical Systems," Ph.D. Thesis, Stanford University, March 2005.

[Kuperberg 2005] Kuperberg, G., Aaronson, S., "Complexity Zoo," http://qwiki.caltech.edu/wiki/Complexity_Zoo, 2005.

[Lakoff 1987] Lakoff, G., Women, Fire, and Dangerous Things, University of Chicago Press, 1987.

[Leivant 1994] Leivant, D., "Higher-Order Logic," in Handbook of Logic and Artificial Intelligence and Logic Programming, Volume 2: Deduction Methodologies, Oxford University Press, 1994.

[Libes 1995] Libes, D., Exploring Expect, O'Reilly Media, December 1994.

[Lierler 2003] Lierler, Y., Maratea, M., "Cmodels-2: SAT-Based Answer Set Solver Enhanced to Non-tight Programs," in Logic Programming and Nonmonotonic Reasoning, 2923, Lifschitz, V., Niemelä, I. (eds.), pp. 346-350, November 2003.

[Lifschitz 2002] Lifschitz, V., "Answer set programming and plan generation," Artificial Intelligence, 138:1-2, pp. 39-54, 2002.

[Lin 2004] Lin, F., Zhao, Y., "ASSAT: computing answer sets of a logic program by SAT solvers," Artificial Intelligence 157:1-2, pp. 115-137 (2004).

[Linsky 1994] Linsky, B., Zalta, E., "In Defense of the Simplest Quantified Modal Logic," Philosophical Perspectives, (Logic and Language), 8, pp. 431-458, 1994.

[Lloyd 1984] Lloyd, J.W., Foundations of Logic Programming, Springer-Verlag, 1984 (second, extended edition 1987).

[Lloyd 1991] Lloyd, J., Shepherdson, J., "Partial evaluation in logic programming," Journal of Logic Programming, 11:3-4, pp. 217-242, October/November 1991.

[Lutz 2003] Lutz, M., Ascher, D., Learning Python, Second Edition, O'Reilly Media, December 2003.

[Magnani, 2005] Magnani, L., "Abduction and Hypothesis Withdrawal in Science," http://www.bu.wsu/wcp/Papers/Scie/ScieMGn.htm, 2005.

[Marek 1999] Marek, V., Truszczy, M., "Stable models and an alternative logic programming paradigm," in the Logic Programming Paradigm: a 25-Year Perspective, Apt, K., Marek, V., Truszczynski, M., Warren, D. (eds.), pp. 375-398, Springer-Verlag, 1999.

[Marshall 1996] Marshall, J.B, Hofstadter, D., "Beyond Copycat: Incorporating Self-Watching into a Computer Model of High-Level Perception and Analogy-Making," in Proceedings of the 1996 Midwest Artificial Intelligence and Cognitive Science Conference, Gasser, M. (ed.), http://www.cs.indiana.edu/event/maics96/Proceedings/Marshall/marshall.html, 1996.

[McCarthy 1980] McCarthy, J., "Circumscription - a form of non-monotonic reasoning," Artificial Intelligence 13, pp. 27-39, 1980.

[McCarthy 1986] McCarthy, J., "Applications of circumscription to formalize commonsense reasoning," Artificial Intelligence, 28, pp. 89-116, 1986.

[McCune 1997] McCune, W., Wos, L., "Otter: The CADE-13 competition incarnations," Journal of Automated Reasoning, 18:2, pp. 211-220, 1997.

[McDermott 1982] McDermott, D., "A temporal logic for reasoning about plans and actions," Cognitive Science, 6, pp. 101-155, 1982.

[Minker 2002] Minker, J., Seipel, D., "Disjunctive Logic Programming: A Survey and Assessment," in Computational Logic: Logic Programming and Beyond, Essays in

Honour of Robert A. Kowalski, Part I, Lecture Notes in Computer Science, 2407, Kakas, A., Sadri, F. (eds.), pp. 472-511, Springer, 2002.

[Moore 1985] Moore, R., "Semantical considerations on non-monotonic logic," Artificial Intelligence 25, pp. 75-94, 1985.

[NIST 2005a] U.S. National Institute of Standards and Technology, "Technology Services, Calibration," http://ts.nist.gov/ts/htdocs/230/233/calibrations, 2005.

[Newell 1957] Newell, A., Shaw, J., Simon, H., "Empirical explorations with the logic theory machine," Proceedings of the Western Joint Computation Conference, pp., 218-239, 1957, reprinted in Siekmann, J., Wrightson, G. (eds.), Automation of Reasoning, Classical Papers on Computational Logic, 1, pp, 49-73, Springer, 1983.

[Niemela 1999] Niemela, I., "Logic programs with stable model semantics as a constraint programming paradigm," Annals of Mathematics and Artificial Intelligence, 25, pp. 241-273, 1999.

[Niemela 2000] Niemela, I., Simons, P., Syrjanen, T., "Smodels: A System for Answer Set Programming," Proceedings of the 8th International Workshop on Non-Monotonic Reasoning, April 2000.

[Nuñez 1991] Nuñez, M., "The Use of Background Knowledge in Decision Tree Induction," Machine Learning 6:3, pp. 231-250, 1991.

[Payne 1990] Payne, E., McArthur, R., Developing Expert Systems: A Knowledge Engineer's Handbook for Rules and Objects, Wiley, 1990.

[Peak 2004] Peak, R., Lubell, J., Srinivasan, V., Waterbury, S., "STEP, XML, and UML: Complementary Technologies," Journal of Computing and Information Science in Engineering, pp. 4:4 pp. 379-390, December 2004.

[Pearl 1988] Pearl, J., Probabilistic Reasoning in Intelligent Systems: Networks of Plausible Inference, Morgan Kaufmann, September, 1988.

[Peng 1990] Peng, Y., Reggia, J., Abductive Inference Models for Diagnostic Problem Solving, Springer Verlag, 1990.

[Poole 1988] Poole, D., "A logical framework for default reasoning," Artificial Intelligence 36, pp. 27-47, 1988.

[Pople, 1973] Pople, H.E., "On the mechanization of abductive logic," Proceedings of the International Joint Conference on Artificial Intelligence 8, pp. 147-152, 1973.

[Protégé] The Protégé Ontology Editor and Knowledge Acquisition System, http://protege.stanford.edu, 2005

[Pryzymusinski 1990] Pryzymusinski, T., "Well-Founded Semantics Coincides With Three-Valued Stable Semantics," Fundamenta Informaticae, 13:4, pp. 445-463, 1990.

[Putnam 1989] Putnam, H., Realism and Reason: Philosophical Papers, Volume 3, Cambridge University Press, 1989.

[Quinlan 1986] Quinlan, J.R., Induction of Decision Trees, Machine Learning 1, pp. 81-106, 1986.

[Quinlan 1993] Quinlan, J.R., C4.5: Programs for Machine Learning, Morgan Kaufmann series in Machine Learning, 1993.

[Quinlan, 1990] Quinlan, J.R., "Learning logical relations from relations," Machine Learning, 5:3, pp. 239-266, 1990.

[Rao 1997] Rao, P., Sagonas, K., Swift, T.,, Warren, D., Freire, J., "XSB - A System for Efficiently Computing Well Founded Semantics," Lecture Notes In Computer Science, 1265, Proceedings of the 4th International Conference on Logic Programming and Nonmonotonic Reasoning, pp. 431-441, 1997.

[Reich 1995] Reich, Y., "Measuring the Value of Knowledge," International Journal of Human-Computer Studies, 42:1, pp. 3-30, January 1995.

[Reiter 1980] Reiter, R., "A logic for default reasoning," Artificial Intelligence 13, pp. 81-132, 1980.

[Riazanov 2002] Riazanov, A., Voronkov, A., "The Design and Implementation of Vampire," AI Communications, 15:2, pp. 91-110, 2002.

[Ringwood 1989] Ringwood, G., "SLD: a folk acronym?," Association for Computing Machinery, SIGPLAN Notices, 24:5 pp. 71-75, 1989.

[Robinson 2001] Robinson, J., Voronkov, A., (eds.), Handbook of Automated Reasoning, MIT Press, 2001.

[RuleML 2005] Rule Markup Language (RuleML), http://www.ruleml.org, 2005.

[Sagonas 1998] Sagonas, K., SWIFT, T., "An Abstract Machine for Tabled Execution of Fixed-Order Stratified Logic Programs," Association for Computing Machinery Transactions on Programming Languages and Systems, 20:3, pp. 586-634, May 1998.

[Savoy 2005] Savoy, R., "DejaGnu," http://www.gnu.org/software/dejagnu, 2005.

[Selman 1988] Selman, B., Kautz, H., "Hard problems for simple default logics," Artificial Intelligence, 49, pp. 243-279, 1991.

[Shapiro 1981] Shapiro, E.Y., "Inductive inference of theories from facts," Technical Report 192, Computer Science Department, Yale University, 1981.

[Simon 1974] Simon, H., Lea. G., "Problem solving and rule induction: A unified view," in Gregg, L. (ed.), Knowledge and Cognition, pp.105-127, Lawerence Erlbaum, 1974.

[SourceForge 2005] SourceForge, "XSB Home Page," xsb.sourceforge.net, 2005

[Sowa 2003] Sowa, J., Mazumdar, A., "Analogical Reasoning," in Conceptual Structures for Knowledge Creation and Communication, de Moor, A., Lex, D., Ganter, B. (eds.), Lecture Notes in Artificial Intelligence (LNAI) 2746, pp. 16-36., Springer-Verlag, http://www.jfsowa.com/pubs/analog.htm, 2003.

[Stickel 1994] Stickel, M., Waldinger, R., Lowry, M., Pressburger, T., Underwood, I., "Deductive Composition of Astronomical Software from Subroutine Libraries," 12th Annual Conference on Automated Deduction, CADE-12, 1994.

[Stickel 1998] Stickel, M., Waldinger, R., Chaudhri, V., "A Guide to Snark," http://www.ai.sri.com/snark/tutorial/tutorial.html, 1998.

[TPTP 2005] Sutcliffe, G.,, Suttner, C., "The TPTP Problem Library for Automated Theorem Proving," http://www.tptp.org, 2005.

[Techsmith 2005] Techsmith, Inc., http://www.techsmith.com, 2005.

[Thagard 1998] Thagard, P., Verbeurgt, K., "Coherence as constraint satisfaction," Cognitive Science, 22, pp. 1-24, 1998.

[Thagard, 1994] Thagard, P., Shelly, C., "Limitations of Current formal models of Abductive Reasoning," University of Waterloo, http://scholar.google.com/url?sa=U&q=http://www.ags.unisb.de/~konrad/reasoning/limitations-abduction.eps.gz, January 1994.

[Thagard, 1997]Thagard, P, Shelly, C., "Abductive reasoning: Logic, Visual thinking and Coherence," http://cogsci.uwtarelloo.ca/Articles/Pages/abductive.html, 1997.

[Trento 2005] Department of Information & Communication Technology, University of Trento, "ASPLIB: The Answer Set Programming Satisfiability Library," http://dit.unitn.it/~wasp/Benchmarks, 2005.

[Turner 1985] Turner, R., Logics for Artificial Intelligence, Ellis Horwood,1985.

[VanEmden 1976] Van Emden, M., Kowalski, R., "The Semantics of Predicate Logic as a Programming Language," Journal of the Association for Computing Machinery, 23:4, pp. 733-742, October 1976.

[VanGelder 1991] VanGelder, S., Ross, K., Schlipf, J., "The Well-Founded Semantics for General Logic Programs," Journal of the Association for Computing Machinery, 38:3, pp. 620-650, 1991.

[Vardi 1997] Vardi, M., "Why is modal logic so robustly decidable?," in Descriptive Complexity and Finite Models, Immerman, N., and Kolaitis, Ph., (eds.), American Mathematical Society, 1997.

[Vilain 1989] Vilain, M., Kautz, H., van Beek, P., "Constraint propagation algorithms for temporal reasoning:A revised report," in: Readings in Qualitative Reasoning about Physical Systems, Weld, D., de Kleer, J., (eds.), pp. 373-381, Morgan Kaufmann, 1989.

[W3C 2004a] World Wide Web Consortium, "Extensible Markup Language (XML) 1.0 (Third Edition)," http://www.w3.org/TR/REC-xml, February 2004.

[W3C 2004b] World Wide Web Consortium, "OWL Web Ontology Language Overview," http://www.w3.org/TR/owl-features, McGuinness, D., Harmelen, F., (eds.), February 2004.

[W3C 2004c] World Wide Web Consortium, "RDF Semantics," Hayes, P. (ed.), http://www.w3.org/TR/2004/REC-rdf-mt-20040210, February 2004.

[W3C 2004d] World Wide Web Consortium, "Resource Description Framework (RDF): Concepts and Abstract Syntax," Klyne., G., Carroll, J., (eds.), http://www.w3.org/TR/2004/REC-rdf-concepts-20040210, February 2004.

[W3C 2004e] World Wide Web Consortium, "OWL Web Ontology Language Guide," Smith, M., Welty, C., McGuinness, D. (eds.) http://www.w3.org/TR/2004/REC-owl-guide-20040210, February 2004.

[Wang 2005] Wang, K., Zhou, L., "Comparisons and Computation of Well-Founded Semantics for Disjunctive Logic Programs," Association for Computing Machinery, Transactions on Computational Logic (TOCL) 6:2, pp. 295-327, April 2005.

[Wang, 2004] Wang, P., "The limitation of Bayesianism," Research Note, Artificial Intelligence, 158, pp. 97-106, 2004.

[Warren 1977] Warren, D., Pereira, L., Pereira, F., "Prolog - The Language and its Implementation Compared with Lisp," Proceedings of the 1977 symposium on Artificial intelligence and programming languages, pp. 109-115, Association for Computing Machinery Press, 1977.

[Warren 1983] Warren, D., "An Abstract Prolog instruction set," Technical Report 309, SRI International, 1983.

[Welch 2003] Welch, B., Jones, K., Hobbs, J., Practical Programming in Tcl and Tk (4th Edition), Prentice Hall PTR, June 10, 2003.

[Welty 2001] Welty, C., Guarino, N., "Support for Ontological Analysis of Taxonomic Relationships," Journal of Data and Knowledge Engineering, 39:1, pp. 51-74, October, 2001.

[Wielemaker 2002] Wielemaker, J., Anjewierden, A., "An Architecture for Making Object-Oriented Systems Available from Prolog," Social Science Informatics, University of Amsterdam, August 2002.

[Wielemaker 2005] Wielemaker, J., SWI-Prolog 5.5 Reference Manual, Social Science Informatics, University of Amsterdam, http://www.swi-prolog.org, July 2005

[Wikipedia 2005] Wikipedia, "Optical character recognition," http://en.wikipedia.org/wiki/Optical_character_recognition, 2005

[Yahoo 2005] Yahoo, Inc., http://www.yahoo.com, 2005.

[Yampoortam 1993] Yampoortam, E., Allen, J.F., "Performance of Temporal Reasoning System," University of Rochester, Tech Report 9193, 1993.

www.ingramcontent.com/pod-product-compliance
Lightning Source LLC
Chambersburg PA
CBHW081731170526
45167CB00009B/3776